Wild Ride

Wild Ride

A Memoir of I.V. Drips and Rocket Ships

Hayley Arceneaux
WITH SANDRA BARK

CONVERGENT
New York

Published in the United States by Convergent Books, an imprint of Random House, a division of Penguin Random House LLC, New York.

Convergent Books is a registered trademark and its C colophon is a trademark of Penguin Random House LLC.

LIBRARY OF CONGRESS CATALOGING-IN-PUBLICATION DATA
Names: Arceneaux, Hayley, author. | Bark, Sandra, author.
Title: Wild ride / Hayley Arceneaux with Sandra Bark.
Description: New York : Convergent, [2022]
Identifiers: LCCN 2022016154 (print) | LCCN 2022016155 (ebook) | ISBN 9780593443842 (hardcover) | ISBN 9780593443859 (ebook)
Subjects: LCSH: Arceneaux, Hayley. | Women astronauts—United States—Biography. | Physicians' assistants—United States—Biography. | Cancer—Patients—United States—Biography. | Bone substitutes—United States. | Space medicine—United States. | Space flight.
Classification: LCC TL789.85.A73 A3 2022 (print) | LCC TL789.85.A73 (ebook) | DDC 629.450092 [B]—dc23/eng/20220527
LC record available at https://lccn.loc.gov/2022016154
LC ebook record available at https://lccn.loc.gov/2022016155

Printed in the United States of America on acid-free paper

crownpublishing.com

2 4 6 8 9 7 5 3 1

First Edition

Book design by Alexis Capitini

To Mom, Dad, and Hayden
By my side, through it all.

Contents

Part 1: Hopes

Chapter 1: How to Pack for Outer Space 3

Chapter 2: Family Is Everything 13

Chapter 3: The New Normal 23

Chapter 4: Dr. Doom and the Cutting-Edge Prosthesis 31

Chapter 5: Losing Hope, Choosing Hope 42

Chapter 6: Valencia 51

Chapter 7: Doctorita 59

Chapter 8: Saying Goodbye 69

Chapter 9: Full Circle 77

Part 2: Dreams

Chapter 10: The Magic Dragon 85

Chapter 11: Inspirati④n 94

Chapter 12: G-Monster 101

Chapter 13: Upper Limits 109

Chapter 14: Inside the Simulations 121

Chapter 15: Nova 128

Chapter 16: Launch Week 137

Chapter 17: Countdown to Launch 145

Chapter 18: Cold Pizza for Dinner (and Breakfast) 153

Chapter 19: Around the World in Ninety Minutes 162

Chapter 20: Splashdown 171

Conclusion: I'm Still Me 176

Acknowledgments 183

A Note About the Cover Art 187

Part 1

Hopes

So much to do, so much to see.
—Smash Mouth, "All Star"

CHAPTER 1

How to Pack for Outer Space

SEPTEMBER 1, 2021
L minus fourteen days

Once upon a time, there was a girl who rode a Dragon to the stars . . .

It's a quiet evening in Memphis, and I'm getting ready for the trip of a lifetime.

"Scarlett," I say, "in two weeks, I'm going to space."

My beautiful, fluffy, gray Aussiedoodle looks up at me with an expression of love and mild concern on her face. She's not worried about me. She's wondering who will watch her when I'm gone. She's seen me pack before, and she knows that it ends with my leaving for a while.

"The boys are going to take amazing care of you," I tell my dog, knowing she'll be safe and happy with our favorite neighbors.

Her expression relaxes and I continue. She's heard it all since the beginning of this wild year: the Dragon spacecraft, the Falcon 9 rocket, the fact that we're going deeper into space than anyone has been in over twenty years. The two of us are quarantining at home together. SpaceX was clear that anyone who tests positive for COVID-19 will not be boarding the spacecraft, and getting sick at this point is just not an option, because there's no way I'm letting anything keep me from the ride of my life.

I can't wait. There's nothing I love more than traveling to a place I've never been before.

Plus, I hear the views are incredible.

In case you're wondering, no, this is not the beginning of a sci-fi fairy tale.

Forget science fiction. This is science fact, and the fact is that very soon I'll be strapping into a spacecraft with my crew. Our mission: After launch, we'll be 370 miles above the surface of the Earth, orbiting for three days at 17,500 miles an hour before we splash back down in the Atlantic Ocean. For reference, the International Space Station hangs out 250 miles up.

The last time I left the United States, I was going on a medical mission trip to treat patients in Nicaragua. The month before that, I was heading to Morocco to ride camels in the Sahara desert. Packing for an adventure is something I've done before, many times, but deciding what to bring to outer space is nothing like planning for a tour of North Africa. Usually, I'm inspired by reading travel blogs and articles by people who have been there before me, but there are no posts about packing for space (I did look).

Most of what I need will be supplied by the mission. Just a week after I was selected, I was fitted for a sharp white space suit by a woman who used to make superhero costumes for movies. That was nine months ago. Now launch is only days away. I'm no superhero and this isn't a movie, but I'm going to wear the hell out of that suit.

This will be my first liftoff, but as someone who's been to five continents (and counting), I know the rush I get when the plane takes off. From an airplane, which flies barely six miles above Earth, it feels like you can see it all—how the coastline juts out in one place and pulls back in another, the way rivers curve through farmland.

We're going to be higher in the sky than any civilian has ever been. I feel so lucky to get a chance to see the world from that angle. Even as I'll be sharing a 26.7-by-13-foot capsule with three other people for three days.

It's a good thing I have a lot of experience sleeping in hostels.

I will never forget the day I got The Call.

It was January 5, 2021, only nine months earlier, and I had a phone meeting scheduled with St. Jude Children's Research Hospital.

As every St. Jude employee knows, we treat the toughest childhood cancers and pediatric diseases, and because of the fundraising we do, families never receive a bill from St. Jude.

St. Jude is everything to me. It's where I work as a PA (physician assistant) with leukemia and lymphoma patients. It's the place that treated me for bone cancer as a kid. Without St. Jude and the big limb-sparing surgery I had when I was ten years old, I could have lost my leg and my life. During treatment, St. Jude became my home. Twenty years later, it's still my home.

I'll do anything I can for this place. When they call on me, I say yes. And they call me often. I've been giving speeches for St. Jude since I was ten!

The desire to fundraise for St. Jude started the weekend before my first big surgery, when my mom and I were heading home to Louisiana so I could see my friends and family before the operation. On the drive south, I kept thinking about all of the buildings that made up the hospital, how many lights they had to keep on, how much it all cost, and how my treatment didn't cost us a thing. Where did all that money come from?

"Mom," I said from the back seat. "When I get older, I want to raise money for St. Jude."

"You don't have to wait until you get older," Mom said. "You can start now."

What she said struck me. I didn't have to wait. Now was now.

I got really quiet for a while, and then I asked Mom to turn down the radio.

"I know how to raise money for St. Jude," I told her. "This is the speech I will give: 'My name is Hayley Arceneaux. I had cancer when I was ten years old and came to St. Jude Children's Research Hospital. They saved my leg and my life. If I didn't go there, my family would either go bankrupt or have to sell our house, but we didn't because St. Jude was free. The staff there is very nice and there isn't one rude or cruel person. When I'm there, I don't think of St. Jude as a cancer hospital but a place where I'm surrounded in love. When people donate, I am very grateful. If you will, please donate money to help pay for the equipment and research to save children's lives.'"

Mom had tears streaming down her face. I thought that it was because my speech was so powerful. Years later she told me why she'd cried that day. "What I was really thinking about in that moment was you," she said. "I was thinking about how you just had to live through that surgery so that you could give that speech one day."

My surgery was successful, and I started traveling for St. Jude giving that talk. If they asked me to give a speech, I would go, no matter where it was.

The first time I traveled for a speech was to New York City. I loved supporting St. Jude, I got a real kick out of public speaking, and I was obsessed with the travel. I was able to help St. Jude *and* see as much of the United States as I could. I loved experiencing the differences from my home in Louisiana, like the snow in New York City, the incredible mountain ranges in California, and the old architecture and history of Boston. The question I would ask my mom after receiving an invitation was "Well, where is it?"

As I got older, the speech evolved, but it's always had the same message: St. Jude is a place where I'm surrounded by love.

Now, wondering what they wanted to talk to me about on the conference call, I said to myself, *I've never said no to St. Jude, and I'm not going to start now.*

My assumption was that they wanted me to be in a commercial or to give a speech, except that their scheduling email had been super cryptic: "We'd love to talk to you about a unique opportunity." And except for that weird feeling I had in my stomach.

Even stranger was who had reached out. Usually, it's someone from fundraising calling with a request. This time, it was the VP of staff, who had never reached out to me before.

I always trust my gut, but in this case, the knot in my stomach made no sense. I called in to the conference line, that weird feeling still with me. Something was coming. I could feel it. I got on the line and saw that a VP from fundraising was on too.

"Hayley, we want to talk to you about something really big," said the VP of staff.

Then he spoke to the other VP. "Do you want to take this away? Do you want to blast off?"

I mouthed the words to myself: *Blast off?*

She started talking about a fundraising effort they were doing.

Okay, I thought, *this makes sense.*

Then it got weird. She explained that through SpaceX, a private aerospace company, a billionaire named Jared Isaacman was going to space on the first all-civilian mission. One of the goals of the mission was to raise funds for St. Jude. Isaacman had given two of the four seats to St. Jude. One was going to be used as part of the fundraising effort. One was for a St. Jude ambassador.

Then she said something even more unexpected. "We'd love to send you."

I laughed. It was the only natural response. Then I said, "Are you serious?"

They assured me that they were very serious. "Will you consider it?"

"Yes!"

After that, it was like talking about any other trip, except that it was also completely surreal. "Space, wow! For how long?"

"Three days."

"Let me talk to my family and make sure they're not adamantly opposed," I said. "But my answer is yes."

After we hung up, I looked at my hands closely, staring at my palms and my fingers, wondering if I was hallucinating. Had that really happened? My hands were shaking. My whole body was shaking.

I FaceTimed Mom. "You are not going to believe this."

My family had been through so much with me, and they had gotten used to my getting on a plane and going on an adventure whenever I felt like it. This was different. How were they going to feel?

However they felt, I knew I had to go.

"I just got invited to go to space."

"WHAT?"

"It's true," I said.

Her eyes were bright; she looked so excited for me.

"I can't pass up this opportunity," I continued.

"No, you can't," she said. "This is once in a lifetime."

We looked at each other and she said, "Call Hayden."

That's what Mom always says. Any issue, her response is "Call your brother." Hayden is the logical one in the family. He's also an aerospace engineer. He's the one who loves space.

When I was a kid, the summer before I was diagnosed, my family took a vacation to NASA. We went on a tour, saw the facility, and learned all about the astronauts. I was interested. Hayden was captivated. That was why we were there in the first place: because Hayden was obsessed. Some kids want to be astronauts; Hayden wanted to design the rockets. Space has always been

Hayden's thing. My thing was working at St. Jude. The idea of space was . . . I wasn't really sure how to process it, I hadn't even dreamed about it, but I wanted to go.

Hayden answered his cellphone from his desk at work.

"It's not an emergency," I said quickly. My family has seen enough of those to share news and information in a way that doesn't scare the other person to pieces.

"I have to talk to you right now. You need to go outside." I waited until he was ready and then said, "Hayden, I got invited to go to space." I watched his face dissolve into shock while my mom looked on, smiling like crazy. I hadn't asked enough questions on the call with St. Jude, I was realizing. "I really want to do it. What's the risk of death, like, fifty percent?"

"No!" said my brother. "Nowhere near fifty. It's less than one percent."

"Okay, great," I said. "Because I'm going."

"Wait, were you planning to go if it was fifty percent?"

"Yeah," I said.

"I wouldn't have let you go!"

I shrugged. "Well, you couldn't have stopped me."

His wife, Liz, is also an aerospace engineer. We called her next.

"Liz, I got invited to go to space, and I'm going."

"What? That's amazing!"

"Do you think it's safe?" I asked her.

Liz is someone who is kind but is also always honest, something I appreciate. "You'll basically be on top of a bomb," she said.

I asked her if she would go if she could, and she said no. After she realized I was still serious about going, she made sure I knew that I had her support.

I emailed my acceptance that night, along with some more questions. "Are we going to any destinations, like the International Space Station? How many NASA astronauts are going to be on board? Are we going to the moon?"

When I told my brother about my response, he rolled his eyes. "Hayley, that was so stupid. You asked if you were going to the moon?"

"Why is that stupid?"

"We haven't been to the moon in decades."

"How was I supposed to know that?!"

Now I know that the moon is 238,900 miles away, but at the time . . .

The moment the world found out that I was going to space, people started calling me, and they all had questions. Many of them wanted to know if I knew that the trip was dangerous. Yes, I did know. But I was going because I believe that you have to take risks to live a fuller life.

Some of them asked me about my relationship with cancer. I told them what I always tell people: I'm so grateful that I had cancer, because it gave me my zest for life.

When I was younger, survivorship was painted for me in a really negative light. Every treatment came with warnings about future side effects.

"Oh, you're getting this chemo. It could affect your heart."

"Oh, this chemo, it could affect your kidneys."

I had three surgeries on my leg between the ages of ten and fifteen, each time replacing my prosthesis, which is an artificial femur and knee that extends from midthigh to midcalf. Not a simple surgery to endure, and the healing process is even worse. Every couple of years, I needed a new surgery, and each surgery took years to heal from. There were points where I thought, *Am I ever going to be off crutches? Am I ever going to be able to walk?*

Years after those surgeries, I've climbed Machu Picchu. I've ridden camels in Morocco and drunk red wine in Spain and hiked Mt. Rainier. And now I'm going to space. It's all because of my love for life, which is because of my cancer. My cancer is why I'm here being adventurous and wanting to meet new people and explore the world, and beyond. It's why I studied medicine. It's why I said yes to space. I know the value of making the most out of every day, because there were times I didn't know if I would have more days.

I plan on making the most of every day in space too, and the items I've packed are a part of that. I was very thoughtful about what I chose. SpaceX packed our clothes and toiletries, so we just had to pack our mementos. I had three lists going to make sure I didn't forget anything that had been special to me through the years.

I stressed the most about what to bring to commemorate my dad. He died of cancer in 2018, and I still forget that I can't just call him and tell him things. It will be left-handed appreciation day, and he is left-handed, and I'm about to call him to say, "Happy left-hander's day!" when I remember that he's gone.

I hate that I can't tell him that I'm going to space. He would have loved hearing that, and he would have been so surprised. I never got to tell him that I got the job at St. Jude, but he could have predicted that. That was in the stars for me. But space? He never would have guessed this. None of us could have guessed.

Then randomly, a few months before launch a colleague was showing me photos and someone in a picture was wearing a tie that looked familiar. It was a printed St. Jude tie that my dad had owned too.

That tie. It was his favorite.

It was also one of our inside jokes. The tie is covered with drawings of faces of kids and flags. I mean, it's ugly. It's a very ugly tie. It's also a St. Jude tie.

I would always say, "Dad, don't wear that tie."

And he would say, "No, I'm gonna wear it. Because then people will ask me about it, and I get to tell them about St. Jude."

We had that conversation about a million times. When I saw that tie in the photo, that was it. I thought, *I'm going on this mission for St. Jude. He would get such a kick out of this. I'm bringing his tie.*

I just knew in that moment. It was the perfect thing.

It gave me goose bumps. *That's what I'm bringing for Dad.*

I've learned that it takes less than ten minutes to reach space. Those ten minutes are going to be physically uncomfortable, with G-forces pressing us down into our seats, but that's what I've been training for. Over the past nine months, I've been spun around and turned upside down, been deprived of oxygen, and practiced swimming in a motorcycle helmet. As Jared, our commander, said early on, we needed to get comfortable with discomfort. And boy, did we.

I'm the G-Monster now, and I'm not afraid of gravitational forces. Once we reach zero gravity, it will all be worth it. I'm planning to float around in our capsule, eat M&M's like Ms. Pac-Man, and have a long, long look at this planet, the place where I have lived my entire beautiful, mysterious, incredible life so far.

It may be hard to believe while I'm gravity-bound on my bedroom floor, but if there's one thing I've learned in my time on Earth, it's that as long as you keep saying yes, everything is possible.

CHAPTER 2

Family Is Everything

Everyone with cancer, every kid whom I work with at St. Jude, has a diagnosis story. This is mine.

I was in the elementary school parking lot with Mom and Hayden after school, walking to the car. Every day was the same: Mom would drive us all to school, where we were students and she was the school psychologist. She had finished her PhD and gotten the job the summer before my kindergarten year, which was what prompted our move from Baton Rouge to the small, quaint town of St. Francisville.

I loved having her at my school and getting to wave at her when I saw her in the halls. And I loved that my brother and I were able to ride with her to and from school every day.

As we walked toward the car, Mom looked my way. Usually Hayden was behind me, but today I was lagging behind.

"Come on, Hayley," she said.

I tried to keep up, but my knee was really hurting. It ached and throbbed, no matter how hard I tried to ignore it.

"Hayley," called Mom again. "Can you hurry up?"

I tried. "My knee hurts, Mom," I said.

I said it again as I caught up to them.

Mom turned around and watched me walk toward her. "Are you limping?" she said. "You are. Are you okay? I'll look at it when we get home."

"It just hurts," I said again. I climbed into the car beside Hayden and we headed home.

Mom was making pancakes, my favorite after-school treat. I limped into the kitchen, and she turned to stare at me, midflip.

"Let me look at you," she said.

It was an unseasonably warm day in late January, and to celebrate, I had worn shorts. She knelt down to look at my knee and gasped, then stood and turned off the burner. Pancakes were forgotten.

"How long have you had this?"

"What?"

She pointed. Clearly visible beneath the hem of my shorts was an egg-sized lump, just above my knee.

"I don't know," I said. "I don't usually look at my legs."

Mom grabbed the phone from the wall and the phone book and called my pediatrician. It was the end of the day, and the doctor was leaving, but she was promised the first appointment in the morning. My mom couldn't shake the worried look from her face.

We still needed our afternoon snack. Mom turned the burner back on and finished making the pancakes.

In my memory of that afternoon, the pancakes didn't taste as good as they usually did.

At 7:00 A.M. the next morning I went to the pediatrician with both of my parents. My dad and I were lighthearted, making jokes. I was hoping to get an Ace bandage out of it, because that would have been so cool, to get to wear it to school. But Mom was not laughing.

The pediatrician assessed the lump above my knee and said it felt bony.

"Let's get an X-ray," she said.

After the X-ray, the tech gave me a lollipop. From where I was sitting, all in all, it was a good morning, even though the Ace bandage had not materialized.

At that time in our small town, X-ray films were usually sent across the river on a ferryboat to the next town to be read by a radiologist. I was oblivious to the seriousness of the situation, not understanding that a radiologist was not a regular doctor.

Instead of shipping them across the river straightaway, my doctor requested that we bring her the films to review.

We waited in the exam room for what felt like ages before she came back in. "This is what I was suspecting," she said. "Osteosarcoma. Bone cancer."

I didn't know what osteosarcoma was, but I knew the word "cancer." And everyone I had known who had cancer had died.

My parents were crying.

I burst into tears. I kept saying, "I don't want to die. I don't want to die."

I turned to Mom. "God must hate me," I said. Why else would this be happening?

Mom said, "No, Hayley. God loves you."

Years later, Mom would tell me that she'd had a really bad feeling from the moment we arrived. Before I went to get my X-ray, she had pulled the pediatrician into the hall to ask, "Could this be something bad?"

And my doctor had hugged her.

The doctor's hug surprised me years later when Mom told me about it, much as it had my mom when it happened. Our pediatrician was not known for her warmth or a gentle bedside manner. When she stepped forward and put her arms around my mother, Mom had her answer.

I remember walking out of the doctor's office, crying and holding my dad. I saw other kids in the waiting room looking at me and thought they might think I was a wimp who was crying over getting a shot or something. I wanted to tell them that this was not the case! But announcing that I had cancer didn't feel right either.

It was late January, Mardi Gras season in Louisiana, a festive time widely celebrated by eating king cake, a ring of pastry covered in colorful frosting—purple, green, and yellow—glazed over with sugar icing, and the best part: a plastic baby doll baked right in. Whoever gets the slice with the baby doll in it is the lucky one. I loved eating that cake. I loved getting the plastic baby.

We ate king cake after we got home from the doctor, and I cried into my serving. It would be many years before I could truly enjoy king cake again.

Before I became the cancer girl, my childhood was idyllic, which I attribute to my wonderful parents. Howard and Colleen were the cutest and the best, starting with the story of how they met.

As they tell it, they were neighbors in Baton Rouge. My dad lived just across the street. I can picture them now, my mom, the PhD student in her twenties, with long brown hair and a wide smile, not much younger than I am now, and my dad, the journalist, a few years older than she was, tall, with a large moustache, which I am told was very stylish for the time.

I grew up hearing their meet-cute story again and again. In Mom's version, there was an older guy whom she would notice staring at her when she went outside. In Dad's version, it's far less creepy: He just happened to notice a pretty girl now and again whenever she was outside her house. The story converges when my mom got a new puppy and took her out for a walk. Dad took this as his opportunity to meet his neighbor.

When he walked up to her, Chloe got so excited that she started dancing around and peed on my dad's foot. According to

Mom, her first words to Dad were, "Oh my gosh, did my dog just pee on you?"

He said no and wiped his shoe on the grass.

I'm here today because a small dog named Chloe couldn't hold it in. That was the beginning of my family.

Until I was ten, I was an active kid who ran and jumped without a thought. When I was seven, my dad, who loved Bruce Lee movies, suggested that we take tae kwon do together.

"Yes!" I said.

Anything he suggested, I wanted to do, because I really wanted to be like my dad.

I had been taking dance and gymnastics, which I really enjoyed, but tae kwon do became my first love. I loved it from the very first class and knew right away that I wanted to go all the way to black belt.

Level by level, Dad and I both moved up the ranks. From white to yellow, then gold, orange, green, blue, purple, red, red with a stripe, brown, brown with a stripe, and one day . . . black. Mom was always in the crowd at my tournaments with my brother, Hayden, filming and taking pictures, giving the occasional wave. I would look out and see her with her camera, smiling, encouraging me as I performed the various forms, sparring with other students and breaking boards with both my fists and my feet.

In order to progress, each time, we had to break a board in half. I loved breaking boards. There's such a rush that comes from putting all your strength and aim into a move and then feeling that thick piece of wood split in two. I would scream a powerful "HUH!" every time I broke a board. It gave me an extra boost of force.

My dad saved a piece of the board I'd broken to earn my brown belt and wrote me a message on it that read, "I'm prouder of you for conquering your fear than for earning your brown belt."

Tae kwon do was a source of so much joy and passion for me, until my dad and I tested for our black belts and only he passed. Despite my confidence, despite all the preparation, I failed. I utterly failed. I was stuck at brown stripe. To my youthful self, this was the biggest disappointment, the most crushing defeat, I had ever known. I had no idea in that moment of the trials that I would soon be facing, the treatments and the surgeries and the years of healing and physical therapy that would follow. All I knew was I was crushed because I was used to being good at things. Failure was new to me.

I look back at that era and see tae kwon do in a whole new way. I wasn't just learning forms. I was learning how to get over disappointment. I was learning perseverance. Those lessons would be so crucial for me as I dealt with the realities of illness. But in the moment, all I cared about was getting that black belt.

It was around the time of testing for my black belt that my leg started to hurt.

"Look at my knees," I would say.

It was just coincidence that my dad had been complaining about exactly the same thing. Soon after he got his black belt, he had arthroscopic surgery on his knee. "My knee hurts!" he would say again and again, staring at his legs.

He would put his knees together for comparison, to show that one was more swollen than the other, and I would peer at them and try to discern what he was showing me. "My knee hurts" and "Look at my knees" became common Dad phrases in our household.

I'd tell my family that my knee hurt too, but unlike Dad's knee, mine showed no difference in size.

"My knee hurts," Dad would say.

"My knee hurts too," I would say.

I complained for a few weeks, pointing at my left knee again and again.

Mom took me to the doctor, but he couldn't see anything wrong. No injury was apparent, which was not a relief.

"I'd feel better if I saw an injury," he said. What he did discern was that my left knee was warmer than the right. Looking back, it was ominously warm.

He recommended Advil and to stay off the leg for a while. Not easy when you're training for your black belt, you know what I mean? I rested, as instructed, although it wasn't easy. With the help of over-the-counter meds, I swallowed the discomfort, finished my training, and scheduled another black belt test.

The second time I tested for my black belt I was nervous, but I worked through it. I felt more confident. I just knew it was my time. And I did it. I finally earned my black belt! Afterward, we had a big celebration. There was even a cake decorated with a picture of a black belt showing "Congratulations!" written below it. My parents presented me with an award to remember the day (as if I could ever forget that victory): a little statue of a person in a tae kwon do uniform wearing a black belt. Except the person in the statue was bald.

"Mom! Dad! Why would you give me this statue of a bald person? I have hair!"

At the time, it just seemed funny.

Then I was diagnosed, and that statue felt like an ugly foreshadowing.

St. Francisville is less than two square miles and has fewer than two thousand residents, a tiny town set just above a curve of the Mississippi River, thirty minutes north of Baton Rouge. It's the kind of town where everyone knows everybody, and that includes your business. Let's just say that news gets around fast.

Maybe an hour after we returned home with the diagnosis, my pastor came over. Brother Babin was a grandfather-like figure with a deep voice and full moustache. He spoke gentle words of encouragement, comforting me.

What I was worrying about wasn't just the idea of cancer. What about *school*?

"I have to go back," I told my parents. "I have perfect attendance!"

I had just been given the biggest, most life-changing news of my life, but I was worried about my perfect attendance record. I see this all the time with my patients today. Kids especially can focus on the simplest details. They like routine, they like to know what to expect. They can have so many big things happening to them and around them, but they just want to know that they will have their favorite toy in their hospital room. Sometimes the simplest details can be the most stabilizing.

I returned to school to maintain my perfect attendance record and told my teachers and classmates what was happening. As I understood it, I had either cancer or a "knot." I was really hoping the lump on my leg would turn out to just be a knot, a nothing. My teachers wore looks of confusion and concern. I didn't mind the extra attention they gave me.

It turned out to be my last day of school for nearly a year.

Meanwhile, my dad was researching my type of cancer online. When he found the website for St. Jude, he called and asked if they would take me. A few days later, my mom sat down with me in the study and told me that I needed to pack.

"We're going to a cancer hospital in Memphis, Tennessee, for a few weeks or so," she said while we both cried. I had just been studying state geography in my fourth-grade class, and I kind of knew where Tennessee was. It felt very far away.

Before we left, I sat with Dad and we looked at a catalog of wigs. Dad assured me that we could find one that would look natural when the time came.

It was a Monday when I started limping. Tuesday, I was diagnosed. By Friday, we were on our way to Memphis. My dad left first, driving so we would have a car. Mom and I flew to "the cancer hospital." I don't remember saying goodbye to Hayden, who was staying with my mom's co-worker. Later, we found his diary from that age, when he was just six, nearly seven. Being the big sister that I am, of course I read it. He wrote, "I wish Hayley didn't have cancer." He spelled many of those words wrong. It was so sweet and made me so sad.

I had flown on a plane a few times before, and now I sat next to Mom, drinking apple juice from a plastic cup, looking out the window, intrigued by the geometric circles and squares of farmland far below.

I was calm until we landed in Memphis, and then I became visibly upset.

"I hate Memphis," I said to Mom while we were carrying our suitcases to the shuttle bus. "It's horrible here. Why would anyone live here?"

Of course, I was just covering up for how frightened I was.

I had no idea what to expect when I got to the hospital, which is something I always keep in mind when I'm working with new patients. The unknown is terrifying, and it's my job to offer information and comfort to people as they process the news they've been given.

I was so young, and I didn't feel ill. Aside from my sore knee, I felt good and normal. Knowing that I was going to a cancer hospital made it all too real. The false hope I had been holding on to that the lump was just some silly "knot" was gone. I knew I had cancer and I had to get treated for it. I knew there would be needles and medicine to take. What else that treatment looked like, I didn't know.

All I wanted was reassurance that I wasn't going to die.

We walked in the doors of St. Jude for the very first time, Mom and I together. As an adult I can only begin to process what that experience was like for my mother. I, of course, had no concept of what this place would mean to me one day.

Mom walked up to the front desk, holding my MRI films.

"I'm here with my daughter, Hay—"

She started to say my name but instead burst into tears. The receptionist came around the front desk and gave her a big hug.

"Don't worry. It's going to be okay. You're part of the St. Jude family now," Ms. Penny said to Mom. "We'll take care of her, and we'll take care of you too."

CHAPTER 3

The New Normal

My first diagnosis accessory turned out to be even better than the Ace bandage I had been imagining just days before: a set of bright red crutches. My new medical team didn't want me to bear weight on my leg because I could fracture my thighbone, thanks to the tumor.

At first, I was unsure on them. I had no experience with crutches—I'd never broken a bone or even sprained an ankle before.

The nurse who was escorting me smiled her encouragement.

"You can do it," she said, beaming at me as I maneuvered these new, shiny, cherry-red things. She had long blond hair, and she said she would be going with me to all of my appointments. There were so many things to do, and she assured me that she would explain each of them to me.

I was grateful to have this nurse at my side, helping us learn the ropes, making me laugh, and making the first day a little less scary. After each appointment I got to choose a prize from the prize box. My favorite was a tube of red lipstick, which I immediately started wearing.

Red lips, red crutches, I was ready for whatever might come my way.

In the coming days, I would become more familiar with the hospital and with all the big words getting thrown at me. It helped when I met with the child life specialist, who was able to translate what was happening into a ten-year-old's language.

To help explain it, she gave me a baby doll.

The procedure I required would insert a central line, a semi-permanent catheter that would go into my chest and into a deep vein. It would be used to administer medication and for blood draws.

We leaned over the doll, and I touched the smooth space on her chest where the catheter would go. Then she helped me carefully insert the central line into the doll's body. We added stitches, just like the ones I would be getting a few days later.

I slept with that doll every night.

St. Jude set my family up in a hotel a few blocks away from the hospital, where we would be staying until we could get into their housing facility. I discovered that the hotel had glass elevators that went up nineteen stories. I loved riding up to the top and seeing the world open up in front of me. Once I had soaked it all in, I'd push *L* for lobby and then do it all over again. Other people got on and off while I just kept looking, enjoying the views of the skyline. Memphis was only a six-hour drive from my town, and we were still right next to the Mississippi River, but everything here seemed big and new. From the top, staring far into the distance, I wondered what else was out there. Maybe we would have some time to explore. Maybe Memphis wasn't so bad after all.

My parents and I were sharing the hotel room. It was comforting to have them close, but we were all also grappling with our own feelings about what was happening. Dad kept trying to lighten the mood. Mom, who usually had a sense of humor, had her serious face on. I was intent on mastering the crutches, and it didn't take me long to get the hang of it. Soon I was swinging on them.

"Look!" I said.

I called them to the door to watch, then took off down the hallway, as fast as I could go. I was thrilled at my mastery, having discovered that I could go faster on crutches than I could on foot, especially with my knee aching. After days of limping, having these pieces of plastic to speed me up was a revelation.

Mom was less thrilled, and her voice carried a warning. "Slow down," she said.

I looked to Dad to tell her to stop worrying, but he had a worried face too.

It made me so angry. I spun around, threw down the crutches, and stomped on my bad leg. I had never been a rebellious child, but there were so many changes going on. It was all too much to hold.

"Hayley!" they both said. Nobody was laughing. Dad looked really upset. If I was trying to get a rise out of them, it had worked. My rebellion didn't last long: A second later I felt guilty at having stressed them out; they were stressed enough already.

I picked the crutches back up and promised to use them at a more reasonable pace.

Biopsy day. I was told that someone from pathology would be present so they could confirm the diagnosis of osteosarcoma in real time. If I woke up with a central line, I would know that I had cancer. No line, the tumor wasn't cancer. But the chance of its not being cancer was slim.

My mom walked me back to the operating room, covered head to toe in medical PPE: a sterile bunny suit, gloves, hairnet, and shoe covers. She was by my side as they were about to put me to sleep.

She leaned over and asked, "Is there anything you want? A special gift after this surgery?"

I didn't even hesitate before replying, "An Expedition car."

She rolled her eyes.

"Okay, you can put her to sleep now," Mom joked to the anesthesiologist.

I woke up from surgery feeling groggy and touched my chest. My fingers met a new bulge in the center where my central line had been placed.

"I have cancer?" I asked. I knew the answer but I wanted confirmation.

Someone said, "Yes."

"Aww, man," I said, and fell headfirst into an unexpected future.

From that moment, I was inpatient every couple of weeks for chemo and whenever I had surprise complications like fevers. As a special gift, my mom promised, instead of a car—I was only ten, after all—I could have something even better: a new puppy. This was something I could look forward to during the new schedule of chemo, blood work, and more chemo.

My chemotherapy cycles were scheduled every three weeks. I would have a block of days receiving multiple kinds of chemotherapy and then a break for a few weeks as my blood counts would fall and then rise again. If it took too long for my blood counts to recover, my next chemo cycle would be delayed. That was often the case.

The chemotherapy treatments hit me hard. I would vomit, nearly nonstop, for days. In the weeks that followed, my appetite was minimal at best and I was weak. My doctor called me the most sensitive-stomached patient she'd ever had.

I distracted myself with dog TV shows and books with pictures of dogs. In between bouts of puking, I would get back into bed, pull a book onto my lap, and lose myself in dreams of the fun I would have with my dog, just as soon as I finished treatment.

As I became close with my nurses, they shared their lives with me. I got to learn all about their pets and their relationship situations. I was always trying to set the single ones up, but unfortunately none of the setups ever worked out. The nurses made me laugh or told me gossip from around the hospital—another perfect distraction.

One of my nurses, Lizzie, was a particular bright spot for me. She was in her twenties, and like me, she came from a small town in the South and was as sweet as pie. She was young but as talented as the veteran nurses, and she was also someone who could throw my sass back at me. I loved her and I always hoped she would be my nurse when I was in the hospital. As luck would have it, she usually was. She's one of those nurses who has a huge heart for her patients and is so warm and present, and also really understands the system and what needs to be done. She's an incredible patient advocate. My sickest days were in the hospital, and Lizzie helped clean up more than her fair share of vomit. She made those hospital stays a great deal sunnier.

There was no way to know it back then when everything was so uncertain, but when I turned twenty-one, Lizzie would be there to buy me a beer to celebrate. I think about that with my own patients, how I'll get to do the same thing with some of them one day.

I got a lot of strength from my doctors and nurses. Even though they were interacting with me in such an abnormal environment, they made me feel normal. They always made me believe that I was going to be okay, and that feeling allowed me to dream. I wanted to do what they did. I wanted to help kids going through cancer treatment get better. I told everyone there I would come back to work with them someday, and I think they believed me.

Since I couldn't go home much during treatment, we all fell into a pattern. Dad and Hayden were in St. Francisville all week while Mom and I were in Memphis. I missed my dad, my little brother, Chloe, and my friends back home in Louisiana. I was staying up to date on what my class was learning with the help of a tutor through the St. Jude school. I often said I didn't feel well enough to do my homework. Obviously, the timing was very convenient when it came to doing homework!

Mom and I were getting used to living together in the apartment in St. Jude housing. Even though our apartment had two bedrooms, when I wasn't in the hospital, I often slept in Mom's room because I didn't want to be alone.

Every weekend, Dad and Hayden would drive up to visit. I would be sitting in front of the window of our apartment, waiting for them to arrive for what always felt like hours. Finally, I would see my dad's black sedan pull up. Then I would rush to grab the scooter my parents had given me, which was by the door. Turns out you really need only one good leg to go fast.

I would fly down the hallway as quickly as I could, carried along by my excitement at having my family together again for the weekend.

I missed home, but as time went on, it turned out that I loved living in the St. Jude housing. There was always something to do, like joining the other kids in the common area to play computer games or do arts and crafts. There were volunteers who would come to spend time with us, even a music room where I learned to play piano. Once a week, a music professor from the local college would come to teach me how to play a complicated ballad. I couldn't read music, but I memorized the hand motions and I would practice on the keyboard in my apartment for hours. Before going to the hospital, I would play my song over and over again.

By the end of treatment, I had finished learning the whole song.

One of the most difficult parts of going through cancer treatment? Losing my hair. I had known that it was coming ever since I had gotten the cancer diagnosis—I knew people with cancer were bald because of their medications.

It started with strands on my pillowcase. I would get so upset seeing them. It made my cancer seem even more real.

"I think it's time to get your hair cut short," Mom said one day. "It will be easier to manage."

"Okay," I said, so she found a salon nearby.

When we got there, Mom went inside first to tell them about my situation so that they could sanitize the station before I came in.

They were kind and gentle with me. The stylist sat me in her chair, gently let my hair fall, and then said, "I think if we cut it just under your ears, it will be adorable." I had never had short hair before and I loved it.

The new short haircut worked for a while, but whenever I brushed or touched my hair, more would fall out. Mom was constantly cleaning behind me so that I wouldn't realize how much I was losing.

I was at home in St. Francisville for a short visit in between chemo treatments when my hair started falling out in larger clumps. There in my bedroom, I realized that I couldn't take it anymore. It was a real turning point for me. I was the sick kid, my hair was falling out, and I couldn't pretend anymore that it wasn't happening.

My mom came into the room and sat down next to me as I put my hand up to my head. I pulled at the ends of some of the strands. I knew what would happen, and it did—the hair came out easily, entwined in my fingers. I pulled on a larger clump, and then another. Handful by handful I pulled all of my hair out. I cried while I did it, though it didn't hurt.

Finally, I took all of my hair and put it in a plastic bag so that

I could remember what it looked like. I still have that bag to this day. Is that weird? Maybe it is. But I can't just throw it away. Seeing it brings me back to that day in such a real way. Sometimes you want to relive and feel even the difficult moments, because they remind you of how far you've come.

I had been a healthy and active kid up until we found that knot above my knee, and now I was really and truly sick. I looked in the mirror and didn't recognize myself. I was so scared that my dog wouldn't recognize me anymore. I needn't have worried, though, because she ran into the room a moment later and straight to me. Chloe had known me since I was a baby. My hair or lack of it didn't change anything for her. I hugged her, still crying.

"You look beautiful," my mom told me. "You are beautiful."

I didn't believe her. I began to avoid mirrors and would cover my eyes so I didn't have to see my bald head when I walked into the bathroom. After a week of that, I knew I had to face it.

I had to face myself.

I walked into the bathroom and looked into the mirror over the sink. I forced myself to stare at my reflection. I knew I had the support of my family, but I needed to find a way to support myself. I stared and stared, trying to see myself for who I was, trying to gain self-acceptance. Several weeks later, when I got back to Memphis, being at St. Jude with all the other bald kids made it easier. We were all in this new normal together.

What a cancer patient goes through is so hard—physically and emotionally. It can embitter you. Being able to look for the good is vitally important as you're traveling through so that you don't lose yourself along the way. I feel so lucky that I instinctively knew that I had to find ways to be happy instead of giving in to the urge to be miserable.

It's not always easy to see the sweetness in life when the taste in your mouth is sour, but I knew I had to try.

CHAPTER 4

Dr. Doom and the Cutting-Edge Prosthesis

When I first went to the hospital, my understanding of the human body was your basic head, shoulders, knees, and toes. But each day that I was there, my knowledge increased more and more, and soon I was an expert on the leg and its bones: femur, patella, tibia, fibula. As I learned, the human leg is a marvel of bone, muscles, and tendons. The two main bones of the leg are the femur, also known as your thighbone, and the tibia, which runs from knee to ankle. My cancer was in my femur.

My first anatomy teacher was the person tasked with saving my leg, Dr. Neel, whom my family met with early on in my treatment. In order to treat my bone cancer and save my leg, he told us we needed to move quickly. The plan was to replace the affected bone with an internal prosthesis. I would have four rounds of chemotherapy to shrink the tumor, after which Dr. Neel would cut out the bone affected by the tumor and replace it with an internal prosthesis, an artificial femur and knee that extends from midfemur to midtibia, that is, midthigh to midcalf. This prosthesis was cutting-edge technology, and since my growth plate was removed in the surgery but I was still growing, this prosthesis could be expanded without additional surgery. That was a very big deal, because a leg surgery like this takes a long time to heal from. We're talking years, not months.

Dr. Neel, whom I quickly started calling Dr. Doom, was a very good doctor. It wasn't despite his bedside manner but because of it. Dr. Doom wasn't like most adults. He didn't sugarcoat. This mattered because it was so different from my experience with most people after I got cancer. Anyone with cancer knows about the shift in the way people talk to you. If you're a kid especially, adults treat you too delicately. Other kids don't know what to say: They're either too nice or they ignore the whole thing completely, and both are confusing when you're trying to get used to this weird and hopefully temporary new normal.

Dr. Doom didn't play make believe with me or talk in an extra sweet sing-song kind of voice. He just told it straight and honestly. It was terrifying at times, and it was something I appreciated.

Before the surgery, he drew my femur with the tumor and his plan for surgery on the paper covering the exam table.

"Remember, I went to medical school, not art school," he said.

That was obvious!

The surgery carried the risk of several complications, and Dr. Doom listed them all, one by one. It sounded gruesome; I didn't want to be a part of any of it.

I got nervous and started loudly gagging, so my mom recommended that I leave the room to play with the child life specialist. A few weeks after we first operated on that doll, the doll had another surgery: a silver bendy straw that we could put into the doll's leg to learn about my expandable prosthesis.

I woke up after the surgery in the recovery room, and physical therapy started right away. One of the first things I saw postanesthesia was a physical therapist walking in with a machine that would be used to help me rehabilitate my leg postsurgery. The machine worked to continually bend and straighten my leg to improve range of motion. Bend and straighten. Bend and straighten. With every bend, with every straighten, I despised the machine a little bit more.

Physical therapy continued nonetheless. Three times a week, as part of continuing treatment, I would go and see the queen of torture, my physical therapist, Lulie.

"Walk pretty," Lulie would say in her Alabama twang.

I would do so, sometimes cheating and keeping my leg straight as I moved. It was easier to walk without bending my knee, but that left a dramatic limp.

Lulie knew I could do better.

"No, walk pretty," she would say to me, again and again.

Following her directives to walk pretty meant that I was forced to bend my knee with each step. I had less of a limp, but it was a lot more challenging.

"Again," Lulie would say, even when I was tired. She was hard on me, in the way that I knew she really loved me and wanted me to succeed. So many other people were too soft with me during that time, but Lulie knew what I was capable of and how far I could go if I worked hard. PT was not always fun, but I looked forward to my time with Lulie.

Then one day Lulie was out. Instead, her boss conducted my PT lesson. I had thought Lulie was tough, but wow. That session was . . . rough. As a result, Lulie was demoted to the princess of pain, and her boss became the new reigning queen of torture.

Over that year, I progressed from using two crutches to one crutch. Then I let the crutch go, and I found myself walking. I wasn't steady, but I was doing it. And I was doing it on my own. Every time I progressed, I rejoiced. Step after shaky step, each one a milestone and a reason to celebrate.

After I graduated from high school, I mailed Lulie a letter telling her that it was thanks to her that I walked across the graduation stage in high heels.

Throughout that entire year, I was undergoing intensive chemotherapy. I received between forty and fifty blood and platelet transfusions and was hospitalized on twenty different occasions for chemotherapy and infections when my blood counts were low. Those are the stats of my chemo, but that's not what I remember about going through cancer. Some of my very favorite memories were made during that time.

One of the best parts of St. Jude was that I got to meet so many new friends, like Hannah and Katie, who are still two of my best friends to this day. I met them early on in my treatment. Those two were always up for playing pranks on the staff.

"You troublemakers!" the nurses would say when they caught on.

I, the inherent rule follower, would protest whenever I was lumped in with them. "I'm not!" I would say.

Hannah had a different kind of bone cancer than I did. She was a year younger than me, but our birthdays are two days apart, which to a nine- and ten-year-old feels a lot like best-friendship destiny. Hannah was energetic and fun and made me laugh. For instance, she once put a goldfish in her I.V. bag and casually walked by the nurses' station.

Katie, who had osteosarcoma like me, was further along in treatment than I was when we met. Her cancer was in the right leg and mine in the left, so when we went out to eat, it was perfect, because we could share a chair to elevate our legs. Her pranks were different, more direct, like spraying Silly String on the child life specialist. The best part is that Katie is now a child life specialist herself.

Though Hannah and I weren't on the same chemo schedule, we'd somehow find ourselves in the hospital at the same time.

Shout out to God, behind the scenes, orchestrating the small details that made me smile and made my time there that much more manageable. The inpatient floor at the hospital had a little jukebox, and Hannah and I made the most of it. We would drag our I.V. poles and our moms to the inpatient floor library and make flyers in old-school Microsoft Word with big, colorful fonts that read, "Come watch us dance. The Hayley and Hannah Show."

We scanned the CD options, but it was easy for me to choose. I went for "All Star" by Smash Mouth every time.

At night, when nobody was roaming the halls, we put the flyers up all around the hospital. The next morning, the nurses helped us decorate our I.V. poles with papier-mâché flowers. Another patient family gave us both costumes to wear. Mine was a purple velvet long-sleeved shirt with tassels and bell-bottoms.

Showtime, and there was a crowd gathered, including nurses, doctors, other patient families, Mom. I danced to my opening song, then vomited on the sidelines while Hannah was performing. We performed another song together. The crowd cheered loudly.

Years later my mom told me how much those dance shows meant to the spectators. To us, it was good, plain fun. To Mom, it was a celebration of life.

At one performance, she told me, she noticed several of the nurses were crying. She asked the nearest nurse what was wrong.

The nurse gestured to a little girl in a red wagon. She was dying and had refused to leave her room for days, but when she heard about the Hayley and Hannah Show, she wanted to see it.

When I heard that story, it hit me hard, and my memories of the silly little dance shows my friend and I did as kids suddenly glowed with meaning.

When I wasn't inpatient, I was back in housing with Mom, but there were still many appointments I had to go to. As an outpatient, several times a week I'd have lab work or a physical therapy appointment. There was a lot of time spent waiting in waiting rooms, which I wasn't into. I didn't need further reminders that I was a sick kid. So I found ways to spend my waiting time that made me feel like I was part of the team, not just a patient.

One day, I discovered the blood donor room, an area separate from the patient area, where members of the community would come to donate blood. I was hanging out in there one day because an adult I knew was giving blood. I liked it so much that I started going back to thank the donors and hand out cookies.

The staff of the blood donor room were really nice about it and even gave me a job title: "Gratitude Administrator."

They mocked up a badge for me to wear. I was very proud to wear it. As far as I was concerned, I was a legit employee.

I would walk around thanking everyone. "Thank you for donating. If I didn't get blood and platelets, I would shrivel up!"

Another way to spend the time between appointments was to join Ms. Penny at the front desk. She wasn't very tall but she had a real presence in the hospital. All of her announcements were funny and personalized. If a celebrity was coming through, we'd hear Ms. Penny over the intercom in advance: "Ladies, put on your lipstick."

The two of us were very enthusiastic when it came to greeting people. I had my patter down. I would always say, "I'm a St. Jude patient, St. Jude fundraiser, and St. Jude employee."

Twenty years later, that's all still true.

My cancer experience wasn't all about the cancer. Instead, it was about the many people who stepped forward to give my family and me support and strength. My mom and I had joined a national Bible study group before I was diagnosed. A local Memphis family heard of our situation through the group and reached out immediately to say that they would pray for us. They even came to visit me in the hospital. Even better, Mom and I got to go to their house, a much-needed break away.

The woman I would come to know as Nana had a beautiful garden in Arkansas, right outside of Memphis. She had four granddaughters and a whole room dedicated to dress-up clothes. Being with them felt like family, and it still does. Two decades later, I still go and visit with Nana whenever I have time.

Looking back, Mom might have needed their support even more than I did. Parents and caregivers of kids going through cancer treatment are the strongest folks in existence. Not only are they dealing with the grief of their child's diagnosis—the fear of losing them—their whole world is suddenly shifted upside down. They have to quickly get up to speed on new words and new tasks, learning to administer I.V. medications and shots to their kid. They're forced to stay strong through it all, even as their child starts to look different, losing weight, losing hair, and no longer having the same energy they once did.

I think often about what this experience was like for my mom. She was forced into this strange new world at the same time I was, and she offered an almost superhuman amount of strength to me during what was such a scary time for her. As an adult, she has shared with me how hard some of those days were for her, especially when she got really bad news, like when I was in the ICU with septic shock from an infection the doctors were worried they couldn't stop, or when the doctors told her that my tumor wasn't responding well to the chemo and they were considering changing my treatment regimen.

After one of the days when she got more bad news, she was driving me back to our apartment from the hospital. She looked at me in the rearview mirror thinking, *Should I take her to the mall?* She wasn't certain I would have much time left and wanted me to create as many happy memories as I could. I had no idea these were her thoughts, no idea of the torment she was struggling with. Instead, I saw only how focused she was on keeping me calm and happy.

Mom was my rock, my anchor in a sea of change. I don't know how she did it. At the time, I just knew that she was kind and loving and always there when I needed her. Now I see the magnitude of what was placed on her and the grace with which she handled everything, always putting me first, above her own needs.

A positive attitude has a lot to do with recovery, something my doctor told my mom early on in my treatment, and I think that's why so many people around me were trying to make my experience with cancer as positive as it could be. Mom worked hard to keep me upbeat. That wasn't always easy, but with her help, and the help of so many people around me, it was a choice I felt like I could make.

I think about Mom's experience today as I take care of many families who have just heard the word "cancer," sometimes only hours or days before we meet, shock and devastation etched into their faces. I walk them through what their next days, weeks, and years will look like. I encourage them by telling the parents, "You and your child will be best friends after all of this." It's true. My mom and I are best friends, and I think a huge factor in what makes our bond so strong is how much we've been through together. We created so many memories, both good and bad, during that year in treatment.

Now that I'm on the other side, working with these caregivers in a provider position, I have a new, deep appreciation for my mom and her strength, and for all of the people who stepped up to support her while she was focused on supporting me.

Two weeks before I was scheduled to finish treatment, leave St. Jude, and head home, Mom and I decided to go to the Memphis humane society to see the dogs there. That special present she asked me about during my first surgery was finally going to be mine. I can't even count how many images of dogs I had stared at over the months, trying to imagine the puppy that would come home with me.

I had finally decided: I wanted a small, white, fluffy little pup. How would Chloe feel about it? I didn't know, but I pictured her taking my puppy around and showing her the ropes of Arceneaux family life. Unfortunately, Chloe died before I finished my treatment. I was sad, knowing that I was going to miss her. I also felt peaceful. She had lived a happy, full life, and I had so many wonderful memories with her. I was sure she would want me to have a puppy to keep me company, since she couldn't be there.

Mom and I walked into the adoption center at the exact same time as a woman who was holding the cutest white fluff I had ever seen. She said she was bringing him in to be adopted. I didn't need to look any further than the entrance. This was exactly the dog I was looking for! The shelter employee told me that since the pup was so young, I would need to wait two weeks before I could adopt him.

I smiled. Two weeks? Perfect.

My mom had the idea to name my new puppy Cottonball in honor of our new friend Nana and her cotton-farming family, and of course because he resembled a little ball of cotton.

Along with looking forward to getting a puppy at the end of treatment, I had been anticipating my "no mo chemo" party throughout that year. St. Jude has a tradition of throwing a "no mo chemo" party after a patient receives their last dose of chemotherapy. There's even a special song the staff sings at every party.

After my last chemo, I walked into the room, still groggy from the antinausea medication, and was greeted by my favorite staff members. I looked around the room as they sang that special song to me:

Our patients have the cutest S-M-I-L-E.
Our patients have the sweetest H-E-A-R-T.

Oh, we love to see you every day, but now's the time we get to say . . .
Pack up your bags, get out the door, you don't get chemo anymore.

It was hard to believe that it was finally happening, but here I was, covered in confetti and Silly String, surrounded by smiling people I loved.

I wanted to cry, but instead I just smiled and laughed. I couldn't cry yet. There was more to come. I wasn't out of the woods just yet, no matter how much pink Silly String was on my head. I had scans coming up in a couple of weeks. I needed to know the results before I could really let myself feel all of the celebration. I wouldn't get the green light to go home until we had received confirmation that the scans were clear.

So even though I was full of joy, I held back happy tears.

Scan day. I sat on the exam table waiting for my doctor to come into the room, all nerves. This was it. This was the moment. *Am I really done? Can I really go home?* The thoughts jumbled around in my head like my brain was in a washing machine, wet and soapy. This was the most nervous I had been about scans. These had the most riding on them.

It took a long time before my sweet doctor, Jane, came into the room holding a stack of papers.

Dr. Jane sat at her computer and then turned to face me, her expression soft and warm as always.

"Are they clear?" I said. The washing machine in my head rolled and turned, rolled and turned.

She nodded yes.

All of the emotions I had been holding inside of me came out in that moment and I broke down in tears of relief. Eleven months of treatment, all of that physical therapy, all of the chemo, all of the puking, the loss of my beautiful long hair, it was finally over! I was alive and I was healthy. My difficult cancer journey was done. I had made it. I was clear.

Although I felt relief, I had another feeling I hadn't expected to have, not with an outcome like this: sadness. The truth was that as much as I was looking forward to going home, I was scared to leave this place that had welcomed us, cared for us, protected us, and helped me heal. I was safe here. I was going to miss this place and this family of people I loved and enjoyed seeing every day.

But little fluffy Cottonball was waiting for me, and so were my dad and brother, and so was the rest of my life.

CHAPTER 5

Losing Hope, Choosing Hope

I headed back home to St. Francisville with a lot of hope in my heart. It was going to be a good year, I just knew it. It was such a triumph to finish treatment, to see those clear scans, and I wish I could tell you that everything from there on out just got better and better. Spoiler alert: It didn't.

Nobody knew what to do with me now. I was treated very delicately by nearly every adult I encountered. They constantly asked me how I was feeling and if I needed anything. They would give me gifts. I could do no wrong in their eyes, and they went out of their way to protect me. I missed my St. Jude nurses, who didn't "Bubble Wrap" me. I missed people who just made me feel like I was normal.

If the adults were extra soft and kind, the kids my own age were basically the opposite.

For example, the biggest kid in school—the classic scripted male bully—zeroed in on me in the hallway a few weeks after I got back to school.

"The only reason people are nice to you and like you is because you had cancer," he said, looming over me.

That little shit! my adult self thinks now. But the child I was in that moment didn't see it that way. I went to my mom's office and cried.

"Do people like me only because I had cancer?" I asked my mom, sobbing.

In Memphis, I had been surrounded by kindness and understanding. I had belonged. I had been loved. This was the complete opposite of all the support I'd had during treatment.

It wasn't just the bully. In my time away, the friends I had before treatment had moved on and made new friends. They didn't seem interested in me anymore. I tried to connect with the girls in my class. I remember one day during recess, one of them brought up Disney World. I was happy to join the conversation; finally we were talking about something we all had experience with. "I loved the—" I started to say, excited to share my love of roller coasters. What I was about to say dried up in my mouth as I watched them quickly move away from me, walking so fast that I couldn't keep up. I stood rooted to the spot and stared at their backs, understanding that this wasn't an oversight. They had ditched me. Not one of them looked back to see if I was coming. I felt like such a loser.

To say that going back to school was a tough adjustment is a massive understatement. It was draining emotionally and exhausting physically. Unfortunately, the one adult in my life who didn't coddle me was the one from whom I could have used some understanding: my grade five teacher.

She didn't seem to care that I had missed out on the first half of the year and was very late to the game with what they were learning. Especially at the beginning, I had trouble keeping up with my classmates. Everyone else would be joking around and running off to their next class, and I'd be totally fatigued. Her class was the first and only one in which I ever got a D on a test.

At the beginning, there were many days when I was too tired

to make it through the full day. I'd ask to be excused and go down the hall to my mother's office. "Mom?" I'd say, standing in the doorway. She would look up knowingly and give me her soothing smile. It was so good having her there in the building in those moments. She was very gentle with me and understood what I was going through. She knew how badly I wanted to just feel better, to be as fast as the other kids again, to have my usual flair for the dramatic and funny.

It took time for me to accept that finishing treatment didn't mean that I was finished with healing. I was quickly learning that my life wouldn't go back to normal as fast as my hair would grow back.

Regaining my strength and freedom and sense of normalcy, I was learning, was going to be an uphill battle.

When the school year ended and summer began, I finally had something to look forward to, something that renewed my hope that things could get better: Camp Horizon. I couldn't have been any happier to go to the summer sleepaway camp. It was made for people like me. And just like St. Jude, it was free.

Mom and I had been collecting the items on the packing list for weeks. I had my toothbrush and toothpaste, my camera, and all of my favorite T-shirts packed into my blue suitcase, which waited in the corner, as eager to go as I was.

Finally, it was mid-June. Mom and I packed up the car, and she drove me up to Memphis, our familiar route, but this time we were going to Katie's house. I would ride the rest of the way to Camp Horizon with Katie and her family.

Camp Horizon is a camp for kids who have or had cancer, set in beautiful rolling hills and woods just outside of Nashville, Tennessee. It had a swimming pool at one end, a babbling stream flowing into a pond in the middle, cabins throughout, and a large fire circle with logs piled into the firepit right in the middle of camp. This was the first of thirteen summers that I would spend there (and counting).

It looked like heaven. As I would soon find out, since there was no air-conditioning at camp, it was actually as hot as Hades. But it is still one of my favorite places on Earth.

Our days were full of swimming, arts and crafts, dance, and seeing friends. Each night we gathered at the fire circle. While the burning logs spit out sparks and the flames illuminated our faces, we sang the Camp Horizon song written by a former camper who had died of cancer: "Beyond the horizon, there's hope for everyone."

To my great delight, my favorite St. Jude nurse, Lizzie, was the camp nurse that summer, cracking jokes and making me laugh as usual. And best of all? Both Katie and Hannah were there. As usual, they pulled me into their pranks and we got up to all the usual high jinks (hello, Silly String). The staff usually let us, pretending they didn't know what we were up to.

The value of Camp Horizon went beyond just having fun. It was a place where I could comfortably share feelings that only friends who had cancer truly understood. At home I was self-conscious of my scars; at camp I was able to be around people who had matching ones. Slowly, I was figuring out how to be myself again.

I started sixth grade with hope in my heart. I had regained my day-to-day strength and mobility, and finally being able to do things without so much effort was a relief. While I couldn't do sports or some of the other activities my peers were doing, I was able to work in the library throughout middle school during my PE period. It was a great way to make friends. By the time I turned twelve in December, life was feeling normal again, and I was there for it.

It is not a fact that whenever we find happiness, something

pops up to ruin it. But that winter, it started to feel that way for me. My leg started to ache. And ache. Until the pain began to get in the way of my normal life. I was constantly uncomfortable and frustrated that I wasn't able to do as much anymore. Although I had graduated from physical therapy, they put me back in, but it only made things worse.

We headed to Memphis in April for my regular checkup and had a regular X-ray. Afterward, we waited for the results. One by one, I saw the other patients leave the surgery clinic until every other patient was gone. At the end of the day, only Mom and I remained, which was not a great sign.

Finally, Dr. Doom came in. "Hayley," he said, "I'm afraid I've got some bad news."

He explained that the expanding mechanism of the prosthesis had broken, presumably during the last lengthening procedure.

"What do you mean?" I said. It ached, but I could still walk on it. How could it be broken?

Until that moment, I hadn't worried at all about anything going wrong with the prosthesis. I'd never given a thought to the fact that it could break. I knew my prosthesis would eventually need to be replaced, but I was sure that day would be years away.

"We have to replace the whole thing," he told me.

I was told I would have to have surgery so that they could put in a "permanent," which is what they call a rod that doesn't expand, because at that point I was done growing.

A range of emotions careened through me, none of them good. I was devastated.

Going through surgery the first time was awful. That second surgery was worse than the first. I spent seventh grade—the year I turned thirteen—in physical therapy. Because our town was so small, I had to go to Baton Rouge several times a week for PT after long days of school. Thanks to the increased amount of scar tissue, I had lost range of motion in my knee, which was endlessly frustrating after how hard I had worked to get where I was.

I started at West Feliciana High School the year I turned fifteen. I know there are people for whom high school was absolutely glorious. Another spoiler: I am not one of those people.

In my freshman year I tried out for the dance team and didn't make it. *No matter, I can try again,* I told myself.

Not making the dance team had been within the realm of possibility anyway. What I hadn't considered—even with the breaking of my prosthesis in sixth grade—was that anything could happen to the permanent.

I was in class one day when the teacher said, "Hey, does anyone want to go to the office?"

"I will!" I said.

I stood up to go—and my leg buckled. The pain in my midthigh was unlike anything I had felt before. I could barely walk.

A girl I didn't know very well grabbed my arm and helped me walk to the office. I was touched by her kindness. At this point, since I was now in high school, my mom wasn't at the same school. But thankfully she was at the school next door, so she came over immediately when I called her. Unlike the last time my leg ached, this time I knew why. My prosthesis was broken. I looked at her, tears in my eyes, and said, "I had been doing so well."

Mom took me straight to the pediatric emergency department, where an X-ray confirmed that it was broken. Another setback, and a huge one at that. I sobbed loudly. I knew what it meant to have the prosthesis replaced.

That revision surgery was followed by six months on crutches and two and a half years of physical therapy. And I lost even more range of motion. The pain and rehab were excruciating to get through. They were some of the darkest moments of my life.

I had always relied on my gift for looking forward and imagining the best, but in those moments, that ability was lost to me. I lost all hope. And without hope, recovery was so much harder.

I couldn't even imagine what my life would look like weeks or months later, let alone when I turned eighteen or twenty-one. I had been excited to go to college, looking forward to my future, to the work I wanted to do at St. Jude. But in these raw moments, I worried that it wouldn't happen. I had so many disappointments over the years because of my health. Maybe college was just another dream I would never wake up in, I thought.

I tried, but I could not imagine better days ahead.

A year of recovery, and I was ready to try again. In my sophomore year, I tried out for the cheerleading team. When they posted the new cheerleading roster, my name was not on it. This time I was not as cavalier. I was sad and disappointed. Why did I even bother trying?

I consoled myself with my upcoming speaking trip. That same evening, I flew to New Hampshire for a speaking event for St. Jude, trying not to be sad, still feeling the sting of not making the team.

Traveling to give speeches for St. Jude was a huge source of pride for me. The speech that I had written in the back seat of my mom's car had grown and evolved, I was out there giving it, which felt amazing, and I met the kindest people through those events.

That night, as I gave my speech, I paid attention to how I felt. I realized that these moments made me feel special, like there actually were places where I fit in. My whole being was flooded with a warmth, and I told myself: *This is what you're good at. This is where you're supposed to be.*

I started to feel like maybe things were going to get back on track.

My first two years of high school had their challenges, but academically, they had been easy for me. Not having to study didn't feel like a win; it felt like a problem. I knew I wanted to work at St.

Jude in some capacity, and my thoughts were turning to medicine, which I knew was a challenging career path. I begged my parents to send me to a new school with a more rigorous academic program so that I would be prepared for college.

My junior year, I was accepted to St. Joseph's Academy, an all-girls Catholic school in Baton Rouge, just forty-five minutes away. I had my driver's license and a silver Camry that my parents had gotten for me. I loved that car so much, but it also made me feel a little guilty. Mom and Dad were the kind of parents who drove old cars so they could put the needs and desires of my brother and me first. I recognized what a sacrifice it was for them to send me to that school. Private school tuition was not cheap.

I was determined to show them my appreciation by doing well in school so that I would then do well in college, get into graduate school, and one day work at St. Jude.

I loved my new school. I had wanted a challenge, and I got it: This place was tough. I had to spend hours on homework and studying each night.

This is what you wanted, I reminded myself, and kept studying.

After working and studying harder than I ever had before, I finished my junior year with one B . . . in *religion* class. Insert eye roll.

Switching schools was a good choice for me—and it was also one of the first life choices I made for myself. At ten, my life changed because of something that happened to me. At sixteen, my life changed because I made something happen. I learned the power of figuring out what I needed and going after it. I was also learning to take control of my identity in a new way.

When a new friend at my new school found out I had cancer, I sent her a private Facebook message: "Please don't tell anyone," I wrote.

This was my fresh start, and I didn't want to be labeled as the cancer girl anymore. I just didn't want to talk about it with my peers. I didn't want them to see me through this lens of illness. I felt that they would see it as a weakness, and that wasn't how I felt.

Speaking for St. Jude was different. In that context, having had cancer didn't make me weak. It was my superpower.

Early on in high school, I wore only long shorts because I was self-conscious of my scars. If my shorts slid up my leg when I sat, I would casually drape my hand over my knee to cover the end of my scar with my hand. People probably didn't notice, but I was always very aware of it. I had a complicated relationship with my leg. I would always call it "my bad leg."

As I got older, instead of seeing my cancer as a source of shame, I felt proud to let people know how far I had come. Cancer was a part of my story. Shorter shorts were a part of my wardrobe.

It took years before I understood that this third prosthesis would carry me through the greatest experiences of my life. It took time for me to become the person who wanted to talk widely about her experiences with cancer. It took time to get to that place where I was confident in who I was, where I could say, "I love my scars." I had to heal, and I had to see things in a new light.

The way I see it now, I have two good legs. Do they look different? Yeah, they do. Do I wear short skirts? Yeah, I do! I'm finally so proud of this leg. My leg has worked so hard and overcome so much. It's gotten me to where I need to be.

In the end it doesn't really matter if other people accept us. What counts is if you can learn to accept yourself.

CHAPTER 6

Valencia

I was in college when I really came into my own. I loved being at Southeastern Louisiana University. I loved being in a sorority, with all my best friends just a few steps away.

I met my best friend, Gabrielle Corsentino, during my sophomore year. We became big sis and little sis in our sorority. On the surface we were total opposites, but our souls and energy were the same. She had begrudgingly gone to college after much convincing from her parents, not yet quite sure what she wanted to do with her life.

I would drag her to the library with me. "We have to win best big/little GPA!" I would tell her.

We were absolutely inseparable, and that meant I gained a whole new family through her extended family. The Corsentinos are a group of warm, down-to-earth people who work hard and play hard. They are some of the most welcoming folks I've ever known, and everyone is invited to their weekend house on the river.

Gabrielle's mom quickly became "Momma Debs" to me.

After a beach trip together (and maybe some dancing on the bow of a boat), she nicknamed me "Haylstorm," and it stuck.

I joined every club I could join while I was in school: my sorority (Alpha Sigma Tau), Up 'til Dawn (an organization to raise money for St. Jude), the Spanish club, Order of Omega Greek honor society, and so many others. I attended all of the social events and weekend football games. I loved meeting new people and holding different positions: spirit chairman, philanthropy chairman, and resident chaplain at our sorority. If you needed a kick of inspiration, I was your girl.

So many sparkling memories: going out to the bars on Thursday night (the "going out" night in our college town), going on beach trips for spring break, going out to eat and trying to spend less than ten dollars in the process on our college student budget, cutting up in the library after hours of studying together. My friend and I invited our dates to formal with a rap song; my rap name was Arce-no-mercy. I'm especially proud of the awards I was given through the years at our sorority formals: Worst Driver, Worst Karaoke Singer, and Most Likely to Die First in a Horror Movie. In a gracious twist, they did round it out with Most Trendy. (After graduating from my all-girls Catholic school and from the plaid skirts that came with that uniform, I loved dressing up. I would get especially dressy on test days to give myself an extra boost of confidence.)

These were my soul mates, those best-friendships that you know are lifelong. I spent so many hours sitting on the overstuffed couches of the sorority house and watching movies with the other girls. I learned about drinking games. I learned that you should *not* get the blue drink from the local bar.

These relationships showed me that I could be my silly self and still be loved and accepted.

But college certainly wasn't all about going out and movie nights and hanging out. I was also taking some pretty tough courses. My biology classes and the rest of the science prerequisites for PA school were the most challenging classes I had. (Until I took an online music theory class. I was totally lost in that class.) I had one goal ever present in mind: No matter what, I had to get a good GPA to get into PA school so that I could fulfill my dream and work at St. Jude. This meant that many nights I would tear myself away from hanging with friends in the sorority house parlor and force myself to go to the library to study.

I was obsessed with the idea of working at St. Jude. So much so that I spent two summers of college interning for them, first on the fundraising side and then for the research department, in the infectious diseases lab.

The summer after my freshman year, while I was interning in the fundraising department at St. Jude, I befriended a guy named Luis who was getting treatment there. Luis and I had so much in common that we couldn't help but become friends. We were just about the same age, had the same cancer, had the same type of prosthesis, and had the same beloved Dr. Doom.

I was minoring in Spanish, so I knew some vocabulary and could speak in basic sentences, but I learned so much more from spending time with Luis's family, the Aguilars. I would join in for their family game nights. Our favorite was Uno, which didn't require complex language to play together. Luis liked to play guitar and sing, and he was always thinking about the younger kids in the oncology clinic back in his home city of Tegucigalpa, Honduras, and would collect toys to take to them. He was a special person, always thinking of others, and I felt so grateful to know him.

The Aguilars had come to the United States from Honduras for Luis's treatment. That summer, I saw their struggles firsthand as they navigated cancer treatment in this country. Through their experience, I saw what life was like for a family working through all of the confusing information in their second language.

I learned something important from my relationship with the Aguilars: Seeing how much harder something already so challenging was with this language gap gave me a new idea about how I could help. Along with studying science, I could make language a priority so that I would be able to work with Spanish-speaking patient families and help them feel more comfortable by treating them in their native language.

When I got back to school, my experience with Luis and his parents inspired me to change my major to Spanish, along with taking my science requirements for PA school.

My Spanish classes quickly became my favorite courses. The Spanish degree program was small, so I got to know my classmates well over the years. Spanish challenged me in a different way. By studying it, I felt a greater and more immediate reward. I could communicate with someone else in what felt like a secret language. I could speak to people I normally wouldn't have been able to communicate with. The best part was our group trip to Valencia, Spain, the next summer. I loved it so much and vowed to return.

With Spanish, it felt like my whole world was opening up.

In my senior year, I wanted to take my Spanish out for another try in the real world, so I planned my first solo trip abroad, back to Valencia.

My mom, who had been to Europe in high school, encouraged me to go. Dad agreed it was a great idea. He had spent his high school years in Greece, spending three years learning the language because he loved Greece's history—and the scuba diving. They both believed in the importance of travel, so much so that they went ahead and bought that plane ticket to Spain for me. I appreciated their support so much.

I was ridiculously excited for that trip. I felt so brave about traveling by myself, so ready to conquer the world. Up until the minute I had to say goodbye to Dad at the airport. It was way harder than I'd anticipated. I reluctantly pulled myself out of his embrace, pushed myself into the line of travelers, then stepped forward into my future. Soon I sat there on the plane wondering what had possessed me to decide to leave behind everyone and everything I knew.

I lived in Valencia for fall semester with a host family: Eva, a single mom, and her two middle school–aged daughters, Paula and Marta. Eva would often do small, gracious things for me. I wasn't an early riser (still am not), but she would make sandwiches for my lunch every day, wrap them up, and leave them outside my bedroom door. Every morning I would open my door to fresh sandwiches with Manchego cheese, *jamón,* and olive oil. How spoiled was I? Eva was giving Dad a run for his money.

I was an extremely picky eater before I went to Spain, so Eva tried to make foods I liked but also encouraged me to try new things. I ate salad and fish for the first time in Spain. Yes, it's true that I had never tried regular green salad or even fish before I went abroad.

Often I didn't know what the foods were.

"*Queso de cabra* is good!" I would say.

"That's goat cheese," Eva would explain.

Goat? I would think. *Glad I didn't know that!*

Everything in Spain was just a little bit different, from the Spanish-language signs to the way you flushed the toilet by pressing a button instead of pulling a lever. Speaking in my second language, I had to think so much more to order a meal, buy stamps, or simply get around. I didn't have my car, so I was forced to learn how to take public transportation. I had never tried reading a subway map before in my life. In my small town back home, we didn't have any public transportation, but in Valencia, I learned to enjoy the efficiency of the metro.

For a twenty-one-year-old American sorority girl, living in a completely new culture was not always easy. My host family was so nice, but I was not always met with the same level of kindness and patience from others as I fumbled along learning their language. A few days after I arrived in Spain, I was in Barcelona, trying to meet up with a friend from the United States who was in town. He was supposed to be waiting at a particular subway stop, and I was trying to navigate the Barcelona subway system.

I went to the station desk and asked the worker there in Spanish, "Excuse me, can you help me?" She fiercely said no and told me I needed to go somewhere else, then shooed me away angrily.

As I turned away, I heard her say to another worker, "UGH! *Americanos.*"

I burst into tears right there in the middle of the station. She didn't know anything about me! Was she judging me by my nationality? And not helping me because of it? I was trying so very hard to speak her language, and it wasn't easy!

After eventually figuring out the subway system and arriving at my stop, I stumbled upon a random café and went in. There I saw a tall, gray-haired man on his laptop sitting in the corner. I went over to him and asked for the Wi-Fi password. He responded to me in English and asked if I was okay.

"NO!" I said.

A floodgate was opened, and it was a long time before I could stop crying to this sweet man. I had never before been treated like I had by that subway worker. The thought of being treated differently for things I couldn't control was all too upsetting to me.

I had similar though luckily less frequent occurrences during my semester in Valencia. It gave me a whole new appreciation for what people who speak English as a second language endure in the United States. I appreciated their courage in such a new and deep way. It gave me a newfound empathy for them. I was determined that once I got back home, I was going to do anything I could to make them feel welcome, especially in my work as a PA.

Over time, I found my way in Spain. I learned that I loved the cappuccinos that cost one euro at the coffee shop in the plaza near home. We'd go out for tapas nights, which were a fun way to try a variety of new foods, like paella, Spanish tortilla, and *patatas bravas*. Top it off with a glass of sangria from the local bar, and you're set. Before I left, I even learned how to make Spanish tortilla—a scrumptious potato and egg dish made on the stovetop—from Eva. I still order sangria and paella whenever I see them on a menu; the taste transports me right back to my time in Spain.

I took three courses during the semester in Valencia: a Spanish-language class for study-abroad students and two with regular local college students. I was the only study-abroad student in both of those regular classes, and wow, those Spanish teachers spoke fast. The beauty of it? I received only pass/fail grades. It was so freeing not to have to keep my GPA up. I just had to pass, and passing was getting 50 percent.

So, yeah, I missed some Friday classes for long-weekend trips to different countries. No regrets. I met other exchange students and we traveled together. Traveling with them helped me to see such a special beauty in other cultures and appreciate in a whole new way people who were different from me. We were a good group. Between our collective languages of English, Spanish, French, and German, we could usually get around fairly easily. All of the exchange students were working to learn Spanish, and we would force ourselves to speak it together, though we cheated pretty often. I went to twelve European countries that semester. I tried to pick up a bit of the language in each one, but it turned out I wasn't great at other languages besides Spanish.

I was made for travel. What I wasn't quite made for was the Valencian "going out" scene, which often started at the bars at 2:00 A.M. and didn't end until 7:00 A.M. I remember one night staying out as late as I could, dancing to EDM under flashing club lights, not caring about drinks being spilled on me or strangers pushing into me, before leaving "early" at 4:30 A.M. I was twenty-one, living in Europe, and having the time of my life. That semester, I saw myself become more outgoing and confident, more open-minded and less judgmental. I eased into the go-with-the-flow attitude because I started to see that things usually had a way of working out.

In Spain, everyone was always telling me to calm down: "No pasa nada." "Tranquila." It took me a while to adapt to Valencia's laid-back lifestyle. Businesses closed during siesta? "No pasa nada." Nothing starts on time? "Tranquila." They must have been getting something right, because the people certainly seemed less stressed.

Near the end of my semester in Spain, I smashed my finger in one of the tall wooden antique doors in the historic center of Valencia. My friends took me to the local emergency room so I could find out if it was broken. When the doctor came in, I said, "I'm scared," and started crying. I knew at the time that a smooshed finger (which, luckily, wasn't broken) was no big deal, but being in a hospital in a different country without my family was frightening. I had a complicated relationship with hospitals after all I'd been through with my cancer treatments.

When I later worked as a PA in the emergency department before I got my job at St. Jude, there were many Spanish-speaking patients who needed my help. I would remember how scared I'd felt during my own ER visit in Valencia and realize that many of those patients were dealing with much scarier things than a finger injury. I always offered them compassion. I wanted them to feel seen, respected, and cared for, and I wanted to take away any fears I could.

I still remind myself of what I learned in Spain when I feel unsure or worried, or when I'm running late to the airport.

Tranquila, Hayley, I tell myself. *No pasa nada.*

CHAPTER 7

Doctorita

I've known since I was a kid that I wanted to work in the medical field and that I wanted to work with kids and their families at St. Jude. That's why I majored in Spanish and studied science. It was in college that I settled on the plan of becoming a physician assistant. I had become familiar with various hospital roles during treatment, and I was always impressed with the role of the PAs on the team. As a PA, I would be able to have a great deal of patient interaction and also have the responsibility of making medical decisions.

I wish I could tell you that I loved PA school as much as I loved college. Steel yourself. I did not love it. PA school is so hard and so fast, like a compressed medical school in two years instead of four.

I spent my nights up late studying and spent weekends in the library. I never forgot my motivation for all of the hard work. I wanted to work with kids with cancer, and nothing was going to stop me. I knew that the only way to achieve that goal was by working hard and not giving up. I was always the one with my nose in a book, always studying, always trying my hardest. It had paid off before, and I had to trust that it would pay off this time too.

I did not really enjoy the fact that all I did throughout PA school was study, but I tolerated it.

I was so happy and relieved when clinical rotations began and I got to interact with patients, show them compassion, and learn from listening to them. That is, until I was about to start my emergency department rotation. I was dreading that experience so much. I was sure I was going to hate the ED.

The morning I was supposed to begin, I called Mom and told her that I wished I didn't have to do it at all.

Turns out I was wrong. It was nothing like what I expected. I loved it! And I was learning that I couldn't always predict what experiences would be best for me.

Throughout my ED rotation, patients would come in with various problems that felt like a puzzle I could solve. I would meet with them and listen to their complaints, order diagnostic tests, quickly obtain and interpret the results, and help them. And it all happened within a few hours, because the ED is so fast-paced.

Expectations and reality can often be very different. Learning this has helped me become more open-minded through the years. You just never know what will work for you unless you try.

As PA school was nearing an end, my chance to make all my hard work worth it presented itself. There were PA jobs available at St. Jude, so I excitedly applied for two of them. I had a wonderful time writing a heartfelt cover letter telling them about how a job at St. Jude was my dream. I remember feeling a huge sense of confidence, like *of course* this was going to work out. This was my plan, right?

I didn't get an interview. When I saw that my application had been passed over without my even being called in for a meeting, I felt like I'd been punched in the gut. It was a disappointment I wasn't ready for.

I had pinned so much on getting the job at St. Jude after I graduated, and I hadn't even been considered for the positions. I was brutally let down and had a good crying session. Then I moved on. I told myself that the timing didn't quite feel right anyway. When I gained more experience, I would apply again to St. Jude. I knew my time would come.

Dealing with disappointments is something kids with cancer get used to. When I started treatment, I was advised not to make any plans because fevers come out of nowhere or clinic days are made longer when labs that day show a blood transfusion is necessary. Of course, we do make plans, because we need things to look forward to, but we know they can always evaporate.

When I was a ten-year-old in the hospital, knowing that other people were having fun while I was not was really hard for me. My family tried not to let me know when Hayden was doing fun things while I couldn't, like when he got to see *Beauty and the Beast* in the theater. If I did find out, I would get jealous, then feel sorry for myself. Years later I'm still finding out about all the fun he had while I was inpatient. My mindset is different now, because I realize how hard it must have been for Hayden too when I was in the hospital, and my parents were probably trying to make things as normal for him as possible.

During the start of the pandemic in 2020, I saw a lot of my friends suddenly wrestling with this new layer of not knowing. They were so used to doing what they had planned to do when they had planned to do it. Suddenly the whole world was subjected to this idea that what you plan just might not happen. It's not easy to absorb that truth, especially when you've been accustomed to writing things in a calendar and having them happen.

Kids like me were forced to learn this lesson early on. One

time, while I was inpatient, I found out that Hayden was out in the free world having fun at a water park with our cousin Lauren. I complained to my mom, saying, "It's just not fair."

A nurse practitioner, who was known for being a little rougher around the edges, happened to be in the room. She turned to me and said, "Well, life's not fair, kid, especially when you've got cancer."

My mom and I looked at each other, stunned. She left the room, and we burst out laughing at her bluntness. On another level, I knew that what she was saying was true.

I thought of her sometimes as I continued to face disappointments through the years, often related to my leg. My Make-A-Wish was delayed when my prosthesis broke the first time. Getting my driver's license got delayed when my prosthesis broke the second time. I didn't make cheerleading because I couldn't jump well.

When I have a disappointment in life, the biggest thing I try to determine is whether it's something I can control or not. Not being able to control unfortunate situations helps me to accept them more easily. Basically, it is what it is.

I do think it's also very important to let yourself mourn. When I talk with people who are recently diagnosed with cancer, they are often quick to assure me they are keeping a good attitude.

"That's great," I tell them. "But you can still let yourself mourn too."

From big to small disappointments, you have to allow the moment where you release the ideal you had in your head. A cry and a venting session can help to get it out so you can move on to look for the positive and hold on to it so that bitterness doesn't set in.

In life, there are always alternate paths. I might not have gotten the St. Jude gig, but I did get a great opportunity to take a position at an emergency department in Baton Rouge. I accepted with gratitude. I loved the excitement and energy of the ED. If I needed to gain some experience, where better to do it than in a busy emergency department? I knew working in the ED would be fun, and after the stress of PA school, I was ready for some fun.

It was going to take a few months for the licensing paperwork to process in order for me to start at the ED, so in the meantime, I took a monthlong volunteer position at a rural clinic in Peru.

Mom and Dad escorted me to the airport. It was the weekend, so they were both available. Of course, that just made it harder to say goodbye.

I hate goodbyes, even if it's just for a month, like this trip was.

A line of people snaked around the security gate. I checked for my passport and then pulled my parents into a tight hug. The three of us stood there for a moment.

"I'm sorry," I told them. "I know my adventures make you nervous. Thank you for always supporting me no matter what."

"Have fun," my dad said.

"We're so proud of you," said Mom. "Be safe."

I moved to the front of the line, showed the guard my passport and my ticket, and was on my way. An hour later, I buckled myself into my window seat. Once again, I was by myself, going to a country where I didn't know anyone.

I stared out the window at the land far below and wondered, *What am I getting myself into?*

Fourteen hours later, I landed at the airport in Cuzco, Peru, and was taken to volunteer headquarters, where we were given information about our host families. I was placed with a roommate, a nurse from Australia named Renee, who was going to be volunteering with me in the same rural medical clinic for the same amount of time. Perfect!

Our host family in Cuzco was a retired older couple who enjoyed hosting our program's volunteers and used it as supplemental income. My host mom was always feeding me way more than I could eat, offering me huge platefuls of rice, beans, potatoes, and chicken, despite my begging her every day to put less on my plate (I had a smaller appetite at Cuzco's elevation of more than eleven thousand feet). I loved drinking coca tea in the morning to help me get used to the elevation, and foamy, tart pisco sours in the evening. A few years had passed since I was the picky eater in Valencia. I was much more adventurous now, though I still couldn't bring myself to try the local delicacy of guinea pig.

My roommate and I ventured out together to explore the city. Immediately, I fell in love with the historic architecture, the local artisanal markets, and the friendly, welcoming people. I walked for miles each day around Cuzco.

The main plaza of the city center has a large church, Iglesia de la Compañía de Jesús, which dates back to 1571 and rests on top of an old Inca palace. It sits alongside several white and stone buildings that form a square around a green space with flowers, trees, and benches that surround a large Inca fountain set right in the center of it all. It served as the perfect meeting spot every time my friends and I made plans. Women dressed in traditional woven patterned dresses with ponchos and layered skirts would walk around with alpacas to take pictures with tourists. It was a tourist trap, and I'm not ashamed to say I fell into it several times. Turns out I love alpacas.

It felt like every day there were parades through the plaza. People in colorful traditional clothing danced through the streets while musicians played music on traditional flutes. The parades always made me happy because they served as a beautiful display of culture.

Also, I'm from Louisiana, and we love a good parade.

I took pleasure in shopping at the local markets. For one U.S. dollar I bought a handmade ring that fit perfectly, a ring I had no idea I would wind up wearing in space. And I couldn't pass up buying an overnight bag with an alpaca on it. Like I said, I love those alpacas.

We were there to spend the month working at a rural clinic an hour outside of Cuzco, but we also found time to explore and take in new experiences. I took salsa lessons every week with the other volunteers, and we spent weekends going on trips around the country. Machu Picchu was at the top of my bucket list and a big factor in why I chose to volunteer in Peru specifically. Seeing Machu Picchu exceeded my expectations. We went up before dawn. The stone ruins were surrounded by mountains covered in low-lying clouds. When the sun broke through at sunrise, everything was covered in light and shadow and the whole landscape seemed to glow.

Another weekend we rode horses and hiked up Rainbow Mountain. With an elevation of over sixteen thousand feet, it was the highest I had been to date. At the top we were able to take only a few steps before we had to stop and catch our breath. On other trips, we rode four-wheelers to salt mines and ancient ruins. My friends and I stayed with a local family at Lake Titicaca, where they dressed us in cultural clothing and showed us their traditional dances. It felt like every evening the other volunteers and I were doing something fun and exciting, which was very welcome after the grueling years of PA school.

It was a wonderful place to practice speaking and continuing to learn Spanish. Since I was finished with my Spanish classes, it was up to me to keep working hard to improve my language skills, even without the motivation to make an A. The ultimate motivation for me now was seeing how Spanish was opening up the world for me. Through Spanish I could learn things and get to know people I couldn't have otherwise known.

The most unexpected part of learning Spanish was how appreciative native speakers were that I was learning their language. They were often surprised that this fair-skinned, blue-eyed girl was speaking to them in full sentences. Even when I knew I was making tons of errors and stammering on words, I was often told, "Wow, your Spanish is so good!"

That encouragement went a long way in giving me confidence and making me want to keep trying.

It was a little terrifying to be on my own for the first time as a PA, let alone in an unfamiliar place, teaching people about best health practices in my second language. I strongly felt this newfound responsibility of being a healthcare provider and reminded myself that PA school had prepared me well and that I was ready to take on patients of my own.

Each morning my roommate and I would take a bus to a different part of the city and wait for someone to go by in a van yelling, "Chinchero" (the name of the rural town). We would hop into the crowded van with the locals. The views along that hour-long drive were beautiful, with alpaca grazing, sprawling fields, and scenic mountains. But they were also humbling. Poverty was all around us. We passed homes that were on the verge of crumbling. Many homes did not have electricity. Each day reminded me more and more how fortunate I was.

We would study Spanish during the drive to the clinic. My roommate was learning Spanish for the first time, so she was studying the basics, especially numbers, since she would be reporting vital signs. I was given lists of vocabulary specific to Peru to study, as well as additional medical terminology.

The clinic was rustic and lacked many basic resources. Wild dogs would even run through the clinic at times. Even so, I was impressed by the staff's creativity in managing. Seeing them build a homemade wheelchair out of a lawn chair with wheels bolted on was inspiring.

The clinic treated all ages, even delivering babies in the back room. It also had an in-house pharmacy so patients could pick up their medications as soon as I finished seeing them.

After nursing staff (including my roommate) collected intake information and took vitals, they would hand me the patient's chart and I would go into detail about the patient's complaint. The majority of the complaints were basic: medication refills or colds. There were very limited diagnostic capabilities—no blood work or imaging such as X-rays, CTs, or MRIs—so I relied more on making clinical judgments from the histories provided and physical exams. People didn't seem to question the fact that I was clearly speaking to them in my second language, with my strong American accent that I still can't shake. They also didn't seem to mind the long waits to be seen.

Since Peru doesn't have a role similar to a physician assistant, the patients assumed I was one of the doctors. I noticed they would call me "Doctorita," or "little doctor." The nickname made me smile. At age twenty-four, I was still considered little. Even though the patients saw me as small and young, they treated me with the utmost respect, kindness, and patience, even when I would ask them to repeat themselves.

One day a patient came into the clinic and I asked what brought him in. He said, "He pisado un clavo."

"Say that again?" I asked in Spanish.

He repeated it, but I still didn't know what he meant.

He picked up his foot and at the same time it dawned on me, "Oh, you stepped on a nail!" I recalled that I had seen that exact sentence while studying that morning.

He was less amused at my revelation than I was, but glad I could help him out.

For a month, I tended to my patients, took their histories in Spanish, diagnosed their illnesses, and wrote prescriptions for their ailments. One valuable tool I took back home with me was how to treat patients even with limited diagnostic capabilities and medications. But the biggest lesson I took away from that experience? How crucial it is that high-quality healthcare be available to everyone. Those in low- and middle-income countries and communities deserve the same access we receive here in the United States. I had so much gratitude for the healthcare I had received, especially as a cancer survivor with a cutting-edge prosthesis.

Healthcare is a basic human right. Healthcare affects patients and it affects their families. We all deserve access to quality, affordable healthcare, no matter where on Earth we live. It's something I fight for to this day.

CHAPTER 8

Saying Goodbye

I finished up my volunteer work in Peru and headed back to the States, where I was finally able to start my new job in the emergency department. I even volunteered to work all the night shifts.

In between, I was solving health crises . . . or advising people who had inadvertently stuck things in places they didn't belong. Some of it was easily fixed, like when people had lacerations and I would just sew them up, or complaints I could knock out with fluids and a cocktail of medicines. Some of it was heartbreaking, like a poor prognosis we just couldn't fix. I learned something from all of it.

Working at the ED was incredible. The work was fast-paced and satisfying. The schedule was flexible, which meant that it allowed me time for my favorite hobby, travel. I was dating a coworker who was a doctor, as if Shonda Rhimes herself had scripted it. It was a perfect time.

Then, about a year into my new job, my dad started complaining of upper-back pain. I had noticed that he was losing weight rapidly. Dad had been dieting on and off my whole life, but he often cheated with Popsicles, so he didn't usually lose weight on his diets. This time was different. His clothes were hanging off him.

PA school had taught me to see the red flags.

"Dad," I said, "you need to see your doctor. Tell him I said that I'm worried you have cancer."

He hugged me.

"It could be a million different things," he said.

"I know," I said. "But we need to make sure."

I had such a bad feeling, like a pit in my stomach. I just couldn't shake it. Every time I saw how thin he was, I would get the same sinking feeling. He, meanwhile, loved the weight loss and would show off to Mom and me how he could fit into clothes he hadn't worn in years.

He scheduled an MRI, which he ended up having done in my ED. It was the strangest meeting of worlds. I wasn't even there, because I was in Las Vegas at an emergency medicine conference. But while I was away, all I could think about was Dad's MRI. I felt nauseous knowing that it was happening and kept hoping my gut feeling was wrong. While I was sitting in the conference, not paying attention to the speaker onstage, my phone buzzed with a text from my boyfriend: "It's bad. . . ."

I immediately got up, walked out of the lecture, and called him. He told me that the MRI showed metastatic disease of my dad's thoracic spine. I knew what this meant. Dad had little tumors throughout his upper back.

It was confirmation of my biggest fear.

I was a mess. This was the worst conversation I had ever had, the worst news, and I was having it with the wrong person. He wasn't Dad's doctor. The emergency medicine physician and PA who were actually treating my dad were friends of mine. We had worked together for over a year at that point. My boyfriend just happened to be in the ED that day, and after the MRI was done, he got into my dad's chart and texted me.

I hung up and called my mom, assuming she had already heard the news. Dad's physician and PA were in the room giving him and Mom the news. Right when I called, they were discussing the best way to break the news to me. But they didn't get the chance. I felt like I was in a nightmare. I didn't know what to do, but I knew I needed to get home right away. Despite searching, I couldn't find any flights back that evening, so I had to wait until morning.

That evening my dad had additional imaging. It revealed a large tumor on his kidney, which was determined to be the main culprit. He also had tumors in his lymph nodes, lungs, liver, and bones. It was as bad as it could be.

I didn't sleep at all that night. I would go through cycles of being numbly stoic, then sobbing hysterically. The next morning, I flew home, crying the whole way. I was so tired, so devastated. I went straight from the airport to the hospital.

It was so strange to see Dad in a hospital gown, the same gown my patients wore. They were my sick patients, and he was my healthy, strong dad. My brain just couldn't put those two things together.

I kept thinking about how I was being robbed of the future I had always seen for my dad and me. He wouldn't be able to walk me down the aisle or be a grandfather to my children. He wouldn't get to watch me turn into the person I would one day become.

When times are tough, people can surprise you, and not always in good ways. Those you think will be there for you can be more hurtful than helpful. You really get to know someone during those difficult moments.

The week after my dad's diagnosis, I broke up with my boyfriend. We were constantly having disagreements and breaking up and getting back together, so this wasn't very noteworthy at the time.

In the moment, I hadn't been mad at him for getting into my dad's chart and telling me. But Mom was furious. She never forgave him for that, because she felt that news should have come from her and Dad. I do agree with her.

Looking back, I think he wanted the power of telling me first. It took me a while to see that I would be stronger without him, which is one of my great regrets. My goal is to live a life without regrets, but sometimes they can serve a purpose. Now I know better.

In that moment, I still had a lot to learn, apparently, because a few weeks after that breakup, we got back together.

Mom, Hayden, and I were by Dad's side as they tried chemo and radiation. The tumors on his spine were causing severe pain, to the point where it reduced his mobility. Watching him suffer was gut-wrenching. He became dependent on all of us, which was so unlike him, this man who always wanted to be in the caretaker role.

When Dad was diagnosed, I promised to take my parents on an international trip, hoping it would give us something to look forward to during a dark time. That's what the idea of travel does for me, maybe as much as the actual going. Thinking about the amazing places that exist in the world can help elevate you above your circumstances. It can give you hope.

Dad was too sick to go. I needed something joyous to hold on to in the moment.

I thought, *I'm going to get a puppy.*

That was when I got Scarlett. In my family, when someone has cancer, we get puppies. We believe in looking forward, in rejoicing, even when our heart is breaking, even when we're grieving.

A few months later I was in Ireland with some of my best friends on a girls' trip when my mom called me in tears. She told me that Dad's condition had worsened quickly, and the doctor had given him only "days to weeks" to live.

As soon as I hung up the phone, I packed quickly and my friends took me to the train station, where I hugged them goodbye. I knew in that moment that I was going home to watch my father die. I was not a stranger to crying in an airport at that point, and I certainly didn't care.

I asked my boyfriend to pick me up from the airport.

"Umm," he said, "I guess so."

We had been planning a party for the next week, when I was due back. After he dropped me off, I texted him to ask if we could reschedule in light of everything going on.

"RIP party" was his response.

RIP party?! I thought. *What about RIP my FATHER?*

I was by my dad's side for the last week he was on hospice before he passed, along with my mom, my brother, and Liz. My mother was amazing. She was losing her husband and best friend, and still she was our family rock more than ever.

The role of the family facing the loss of a loved one was becoming clear to me. For all of the physical suffering I had undergone, I was totally naïve when it came to this kind of grief. When I was going through treatment, I was very aware that it was *me* who was suffering. Sitting in the hard-backed chair next to my dad's hospital bed, I felt an overwhelming feeling of helplessness and fear. It was how my family must have felt when I was going through treatment. I had never been in the family-member role before, watching a loved one suffer, and I didn't like it.

We were with Dad to the end. We asked him what he thought we would miss most about him, and he said his humor. He was right about that, among so many other things. We played his favorite music, the Beatles and the Beach Boys, in his final days. He was slipping away quickly and peacefully.

We spent his last days having long talks and sharing what was most important.

My dad told us, "The most important things in life are family and showing love. I don't know how much time I have left, but I will continue showing love."

I texted my boyfriend, asking him to come and be with me. He was nowhere to be found, despite having that week off work. He did, however, text me pictures of him drinking beer with his friends.

I texted him again, telling him I was disappointed that he didn't once come to the house to be by my side as my dad was dying. I was in such a vulnerable state in that moment. All I needed was the support of my community, but he had none to offer. Instead of compassion, he attacked me, my job, my attitude, even my pain. His responses were cruel and pointed. "You're playing a part in all of the bad things that happen to you," he texted me.

I knew I couldn't, shouldn't, wouldn't play a part in this relationship anymore, that was for sure. I broke up with him for the last time.

The next morning, my mom came into my room in tears and said, "I think he's gone."

I grabbed my stethoscope and went in to listen to Dad's heart, but I heard no heartbeat. I felt no pulse. He was gone.

A few hours later it was raining while the sun was out. For some reason, I took that as a sign that he was okay.

There's so much shit you have to do when someone dies. Let everyone know, write an obituary, plan a funeral, write a eulogy, make a photo display, pick out a grave site. At the same time, you're really sad and don't want to do any of that. But you get through it somehow.

Dad's funeral was in a beautiful church with beautiful flowers and beautiful music. Hayden and I did Dad's eulogy. Dad was the town journalist, and everyone knew and loved him. He would spotlight small businesses and local programs. When he passed, the whole town came to pay their respects and honor his incredible life.

My now-ex was there too, sitting in the back. We had broken up for the final time eight hours before my dad died. After the funeral, he asked me to walk to his car with him.

I told him no. "You realize we're not getting back together, right?"

He looked shocked.

"You failed the final exam," I said.

Not supporting me when I needed it most and instead showing cruelty was unforgivable. That is not what my dad meant about showing love.

I had always had the best relationship models. My mom was Dad's biggest champion and constant companion, and my dad was so grateful to her, calling her his saint. Before he died, he told us to listen to her and said that he couldn't have imagined a better family or a more wonderful partner.

After the funeral, I told my mom she could choose anywhere in the world to go, and I would plan it. I wanted her to have something to look forward to. She picked Italy and Switzerland, and I took two weeks off work to travel with her. We started in Venice because it was one of my favorite cities when I visited it during my study-abroad semester, and I wanted to introduce it to my mother. We rode in gondolas because in Venice, you just have to.

From Venice, we headed to Cinque Terre, where Hayden and Liz met up with us. I watched from the window with anticipation, waiting to see them walk up the hill to the hotel with their suitcases. I recognized the feeling of anticipation; it was the same feeling I had as I watched from the window in Memphis when I was ten, waiting for Hayden and Dad to arrive.

We had so much fun together. We took a pesto-making class with a view of the colorful, historic buildings on the mountainside overlooking the Mediterranean. We swam at a rocky beach. We kayaked. A local man took us on his boat and we swam in perfectly clear green-blue water. At Lake Como, we explored the towns along the lake and went to the top of one of the mountains to see the full panoramic view.

The best part? We spent so much of our time together just laughing, even when my mom tripped and fell (after we found out she was okay, of course). And we shared so many stories about Dad.

After a week, Hayden and Liz headed back to the United States for work, and Mom and I continued on to Switzerland. We stayed in Bern and Lucerne and took day trips by train to several small towns, where we explored the countryside. We hiked behind a roaring waterfall and drank Swiss beer and ate fondue. The skies and water were perfectly blue.

Mom was so happy, at peace. Dad would have wanted that for her more than anything, for her to feel so loved.

CHAPTER 9

Full Circle

Travel became a coping mechanism for me. When I would get upset or go through something difficult, I'd immediately start planning a new trip. Post–Dad's death, I kept traveling while I worked in the ED.

I did yoga in Tulum (Mexico), rode camels and camped in the Sahara desert in Morocco, hiked on a glacier and bungeed in New Zealand. By this time, I had visited more than twenty countries. Traveling brought me so much joy. It made me feel the most alive. I got to leave stress and day-to-day worries back home and just be there in the moment, enjoying beautiful views and new experiences. Having a trip on the docket gave me so much to daydream about and look forward to. Scanning through travel blogs and Instagram posts, figuring out the best places to see, and traveling to a different country came with a whole new set of challenges that I welcomed: new languages, currency, food, cultural expectations. I wanted to learn it all.

After three years in the ED, I felt in my gut it was my time to get my dream job at St. Jude. At the same time, an opportunity presented itself that was just perfect for me.

This time, I got the interview.

The last time I had walked through the doors of St. Jude, a year and a half earlier, I'd been a patient. Now it was time to become an employee. Walking through the halls as an interviewee, they felt smaller than I remembered. I was nervous, but I tried to put on a confident face.

I interviewed with many people throughout the day, including a few panels.

"This is my dream job," I told them all, and listed all the ways I'd worked toward it through the years.

Mom called as I was driving home. "Hey, birthday girl," she said. "How was it?"

My birthday wasn't until the next day, but in my family, we like to start celebrating early.

"So good," I said. "So good!!"

I don't tend toward overconfidence, but I just felt like it was the right time. Plus, as my mom had said, it was basically my birthday.

Since I prefer experiences over stuff when it comes to gifts, my birthday plans revolved around a friend and a Broadway show. The flight was that evening. All I had to do was pack and chill.

The second I woke up, I checked the St. Jude website.

My application status had changed to "not selected."

I spent the day ugly crying. I was devastated.

How could I not have gotten my *dream* job? The timing was right. I had done everything I was supposed to. St. Jude was where I was meant to be, and I was ready. But here was this big fat "no" staring me in the face, telling me that there was no place for me.

Family and friends kept calling to wish me happy birthday. As they did, I would cry in response.

"No," I kept repeating. "No, I didn't get the job."

My loved ones were as shocked as I was. They knew how long I'd been talking about working at St. Jude, and they had been nearly as excited as I was to know I was interviewing. They offered encouragement and loving words.

Eventually I stopped crying. I had a plane to catch. I had a ticket to a city I loved, it was my twenty-eighth birthday, and life wouldn't wait just because I was disappointed.

Besides, I wasn't giving up. I would find a way.

Disappointment, rinse, repeat.

A week later, out of the blue, I got an email from HR telling me they had a job they thought I'd be perfect for, with the inpatient leukemia and lymphoma team. I applied right away.

I remember flying to Memphis to my second interview in some of the worst turbulence I'd ever been in. Despite it, I felt an overwhelming peace. *I'm not going to die in this storm; I'm on my way to get my dream job* was the thought rolling around in my tranquil head.

From the airport, I called Mom. "Do you remember eighteen years ago when we were here and I said, 'Who would ever want to live in Memphis?' "

"Yes," she said.

"Well," I laughed, "now I'm praying harder than ever that I'll get to live in Memphis."

I had another long day of interviews and panels, but I felt such a spark with this team. I was inspired by the way they spoke about their passion for the job and how they were able to help the patients and families. I felt at home.

I met with Lizzie in the cafeteria before I flew back.

"Good luck," she said. "I believe in you."

I didn't hear anything for about a week, and my nerves were increasing by the day. Then one day I woke up at 5:00 P.M. for my night shift in the ED. I flipped the laptop open, checking again to see if my status had changed.

It had.

I was in.

I ugly cried again, this time with overwhelming joy. Finding out I got my dream job was the happiest moment of my life. Working at St. Jude was all I had ever wanted. The job wasn't just given to me. I worked my ass off for it. All the sacrifices I made to study, the experiences I gained working in the ED, how hard it was to learn Spanish.

All of it was worth it.

It was going to take some time for the St. Jude paperwork to be processed, so in the meantime, I went to Nicaragua. It was a medical mission trip, where we took down supplies and saw patients. I was with other physicians, and together we were able to treat the patients with the medications we'd brought. We also got to explore the beauty of Nicaragua, which was raw, natural, untouched.

We had recently heard news of a virus that was affecting Asia. On the plane ride back, I sorted through my work emails that outlined what the plan would be if it hit us.

Two weeks later we were in lockdown for COVID-19.

I ended up working in the COVID-19 unit of the ED while still waiting for the paperwork to go through for the St. Jude position. I was so nervous about working with this deadly virus that we still didn't know much about. I would play reggaeton music loudly before work in an attempt to not think about what I was about to walk into.

Mom was especially worried. When I told her that I was scheduled in the COVID-19 unit, she loudly screamed, "NO!"

I had never seen her react that way toward anything, not even my risky adventures.

"Mom, I have to," I explained. "If I don't, someone else will."

She worried about my kidneys, which had taken a slight hit from the chemo I'd received years ago.

We still didn't know the risk factors for who would be most affected by the virus, but working with patients was something I had to do; it was my duty.

I ended up loving working in the COVID-19 unit. It was the most fulfilled I had felt working in the ED for those three and a half years. Working the front lines of a global pandemic, I was helping people more than ever. I saw and heard some deeply upsetting things during that time, but I liked working for something bigger than myself.

In late April 2020 I started my first day of work at St. Jude. No patient band on my arm this time. Now I had an employee name badge around my neck. It felt so empowering. In my employee ID photo, my bangs were grown out to a weird length because hair salons were closed during the lockdown, but I had the biggest smile on my face.

I met so many people on my first day and tried to remember their names and half of their faces, the other half covered by a mask. I had this incredible sense of fulfillment, but I was also nervous. I wanted to make a good impression and do a good job. I had been given enormous trust and responsibility, and I had to prove that I deserved to be there.

After Dad's death, I was angrier at cancer than ever before. Cancer took some of the people I've loved most. It's crippled people I love. We had to suffer because of it. Cancer looms over survivors like a constant threat. I've been told since I was young that I'm at a higher risk for getting cancer again than the general public. When I have random aches and pains, getting another cancer diagnosis is always on my mind.

But I also believe that having cancer has been my greatest gift, something I learned over time. However, I would never say that to a child or anyone who has just been diagnosed. Discovering that will be their journey. What I tell my newly diagnosed patients is that cancer will make them who they are and that they're going to be stronger because of it.

Childhood cancer gave my life purpose and gave me a sense of adventure. I have such a love and zest for life that I attribute to my cancer journey. I wouldn't take it back for anything. I appreciate the healthy days so much more than I ever would have before. I love every day that I'm alive. I love getting to experience as much as I can of what life has to offer, because it's not guaranteed, which I know all too well.

It took a while for me to realize it, but I also became a better patient care provider because I went through my dad's cancer. It made me understand the family side of things, how helpless and afraid a patient's loved ones can feel. Now, as I cared for the kids, I knew I was also caring for their families, so I really tried to empower them with information and give them all the time in the world to ask their questions.

Working at St. Jude, I felt that everything was finally fitting together perfectly.

I love the unexpectedness of life. I love that things work out. You have to hold on, because you don't know what great thing can come and change your life.

Anything can happen.

Part 2

Dreams

You'll never know if you don't go (go!)

—Smash Mouth, "All Star"

CHAPTER 10

The Magic Dragon

Working at St. Jude, treating kids as I was, everything made sense. I was exactly where I should be.

My days were full. By 7:00 A.M. I was at work, receiving my list of patients and reviewing their histories, vital signs, lab work, and overnight events. I would go around and talk to the kiddos, hear their complaints, and examine them. At 9:30 A.M., we had multidisciplinary rounds with the attending physician, other PAs/NPs, the hematology/oncology fellow, the nursing team, the pharmacist, and the nutritionist. I presented my patients as well as my plan for the day.

"I'd like to consult neurology on this one and discharge this one," I would say, making sure to take any medication changes and follow-up plans into account. After rounds, I would implement the final plan. When additional labs and imaging were resulted, I discussed the results and subsequent plan with the patients' families.

Throughout the day, more patients were being admitted. Some of these kids with suspected cancers were arriving at St.

Jude for the first time. Doing a diagnostic workup was intensive, and I tried to be as present as possible with the families.

"I know this is overwhelming," I told them, walking them through the process and trying to alleviate their fears.

I don't always share with my patients that I am also a former patient. Our encounters are about them, not me. If I think a patient could benefit from hearing it, I will share with them. "I've been there," I'll say in those cases. With new families, it is a way for me to connect.

I've had mothers (and fathers) break down in tears. I can see it, the shift in their expressions from despair to hope. Someone like their child has made it through to the other side.

Memphis was on lockdown when I moved there. I lived in an apartment near a park, which was great for getting outside, and I enjoyed long walks with Scarlett to occupy our time. I discovered that my perfect week included a Saturday trip to the local farmers market and yoga on Sundays. I loved Memphis culture. The blues. Elvis. The rich history of this city, which has a deep connection to the civil rights movement.

As the city opened up, I connected with people, old and new, from those I'd known since I was a child at St. Jude to colleagues and friends of friends. Memphians are kind and welcoming people. Soon I had a busy social agenda. Most nights, I went out to eat, happy to indulge in Memphis's culinary scene. This is a big city, yet it felt like a small town in that I soon saw people I knew almost everywhere I went.

After a few months in the apartment, Scarlett and I went house shopping. I wanted an old house, because they are full of so much character, but one that had been redone in a modern way. When I saw the white brick house with three bedrooms for my friends to stay in, with its arched walkways and even a cutout where an old phone used to sit, I was instantly in love. Plus, it had a nice backyard for Scarlett.

I'd always known I wanted to live in Memphis, because I wanted to work at St. Jude and Memphis comes with it. I'd always viewed Memphis as my city. Now it truly was.

I was home.

It was at this point, as I was getting adjusted to my life as a St. Jude PA, that the chief of staff at St. Jude emailed me regarding a "unique opportunity."

That was the moment when everything changed again.

These are the kinds of things that remind me that there's Someone behind all this. My life couldn't have worked out like this without Someone setting it in motion. Like Dad always said, "Coincidences are God's way of remaining anonymous." You never know what's going to happen, but you have to hope, try, and say yes.

On January 8, I had my first Zoom call with the man who would be taking me to space. After rounds, I told one of my co-workers that I had to escape for an hour to take a call. I couldn't tell her what it was about, but she picked up on the fact that I had to be impressive on the call. I was visibly nervous.

What if I meet this guy and he doesn't like me and picks someone else to go to space? I had known about the mission for only three days, but my heart was already so set on going.

My co-worker took me to an empty interactive room and chose a waterfall for the screen behind me so I looked like I was in a Costa Rican jungle. I joined the call, and there they were: Jared Isaacman, the guy I had been googling all week, and Kidd Poteet, the man who would be our mission director.

I could tell that Jared was tall and younger than I'd imagined, and he had the most genuine smile. Kidd was grinning too, and their friendly expressions put me at ease.

Jared had been described to me as this billionaire businessman CEO fighter jet pilot; but there talking to me on the screen was a really normal, hilarious guy who started joking around immediately. Kidd was an ex-Thunderbird (U.S. Air Force), and he and Jared seemed to know each other well. Later I would learn that they were great friends who had spent years flying and climbing mountains together.

This was the first time I was hearing the name of the mission, Inspiration4. "Inspiration" because Jared hoped the mission itself and the crew would inspire others. The number "4" was because there would be four crew members and an homage to his company Shift4, a payment-processing company for businesses that he started when he was just sixteen years old.

The mission's goal was to raise $200 million for St. Jude. *Wow!* It was the biggest fundraiser the hospital had ever seen. Jared mentioned that he was donating too. He was nonchalant about it. Later I found out that his donation was $125 million. Everyone always asks if he created this mission because he has a personal relationship with childhood cancer. He doesn't; he just hates childhood cancer and wants to see it eradicated.

Jared outlined the details of the mission. There were still so many unknowns at the time, but he had the basic outline. The amazing thing was that the whole crew would be made up of civilians. Jared had his pilot's license, but he wasn't an astronaut, and he had never trained for or been to space.

I had prepared a list of questions, this time with the help of my brother so that I didn't ask silly questions like whether we were going to the moon. Jared proved to be so approachable and easy to talk to, not the intimidating billionaire I had in mind. I told him how excited I was, and how grateful I was for being chosen. I could tell he was excited too.

He invited me to come out the next week to the SpaceX complex in California, where I would start my medical evaluations and get fitted for my space suit.

"It'll be the most intense set of measurements you'll ever

have taken," he said. "Two hundred and twenty measurements, to be exact."

Then he smiled even bigger. We were going to be able to talk live with St. Jude patients and their families from space. I would be able to bring all of my patients into that moment with me. I thought about how much hope it would give those kids watching it as well as their parents. When I was in treatment and my mom met a healthy adult with a childhood cancer story, she would burst into tears. She didn't cry much when I was going through treatment, and doesn't even now, but she was always very emotional after seeing a pediatric cancer survivor grow up to lead a healthy life. Especially because we didn't get to meet many.

Kidd texted me as soon as the call was over so I would have his number. He said they would pick me up in Memphis the next week to head to SpaceX.

"Pick me up?"

"Private plane," he explained.

That was a new concept for me. Okay, it was all new for me. As you might expect, processing all this information took a minute. Okay, a few minutes. I was still reeling from hearing that I would be fitted for my space suit in just a few days, let alone that I was going to space.

This was getting very real.

I was to meet up with Jared and Kidd in Memphis's small private airport, and I had been told that a film crew would be there to capture the moment because our journey was going to be filmed and made into a documentary.

I worked that morning at St. Jude, ran home, quickly curled my hair and changed out of my scrubs, and made it to the airport just in time. A small plane covered in black and gray swirls touched down.

"That's Jared's plane," someone told me.

I stood just inside the door while a sound man put a mic on me; two cameras were facing me, with a sound boom overhead. As if I needed any more reasons to be nervous in that moment! I didn't comprehend the significance of our mission yet, and I wondered why a documentary was already in the works.

Jared and Kidd walked up, both wearing Inspiration4 vests over T-shirts, jeans, and tennis shoes, what I would come to call the Jared uniform. In my black dress and square-heeled white leather boots, I felt overdressed.

The film crew stood behind us as I got ready to greet the man who would not only change my life but also become such an important person to me.

"Hello!" Jared and Kidd called out, smiling broadly.

"Hey, it's so nice to meet you," I responded, looking up at them.

Jared was even taller than I had imagined, a contrast to my five-foot-two stature. He was so friendly, and yet I felt so awkward. The microphone pinned to my neckline wasn't helping.

We were led to a conference room, where there was a spread of Memphis barbecue waiting for us. The food looked good, but I was nervous and barely touched it. An hour later, it was time to go.

"Nice plane," I told Jared as we climbed the stairs, ready to head to California.

"Even though the design looks like Mike Tyson's face tattoo," he said.

I laughed. "Now I can't unsee it," I said.

Okay, he's cool, I thought. *This is going to be okay.*

The next morning, Kidd drove us from our hotel in LA to Hawthorne, where SpaceX was located. I was impressed by the collection of tall white buildings and the large rocket booster looming over the front door. It was incredible to see. Here they were making Dragon capsules like the one I would be taking to space and other rocket parts.

A girl my age, also named Haley, the astronaut operations lead, gave me a tour of the facilities. She was tall and mysterious, and while I had no idea at the time that she would become one of my closest friends, I probably should have guessed.

She walked me through the rocket factory and into the astronaut training room, where I was going to be fitted for my space suit.

"Try these on," someone said, handing me a pile of clothes.

I went into the changing room and slid into a tight black long-sleeve shirt, tights, and socks to wear for the fitting.

Once changed, I was greeted by a team of space suit engineers.

"I'm Maria," said a blond Swedish woman with a tape measure around her neck. By the end of our session, I would find out that before working for SpaceX, Maria had helped create superhero costumes for films. It certainly felt like I was being measured for superhero gear.

The process took hours. My hands and feet were traced, and all the different angles were measured. She wrapped the tape measure around every inch of my arms, legs, and torso and read the numbers aloud, 220 measurements in all. The engineer flanking her documented each figure with diligence, every inch a crucial part of the whole.

Next the team fitted me for my seat in the capsule, taking care to get the width and height of the seat just right to ensure that I'd be comfortable for the long hours preceding liftoff.

After Maria measured every bit of me, she handed me a space suit helmet to try on. I slid it over my head, suddenly in a world separate from everyone around me, all outside sounds muffled. I turned my head from left to right, looking around, getting a feel for it.

Later in the day, Jared joined me in the astronaut training

room. We went and sat in our training capsule, an exact replica of the Dragon capsule that would be our spacecraft.

"This is where we're going to live for three days," he said.

"Home sweet home," I replied.

We were both quiet for the next several minutes as we looked around, studying the structure. I was imagining what it was going to be like to live in this capsule in space. Imagining who else would be joining us. Imagining what the views would look like.

I spent three days at SpaceX on this first trip, and I was back two weeks later to continue my medical evaluations. During my second trip, I had a meeting with our SpaceX mission manager, an attorney, and a big stack of papers. Together we went through the likelihood of every negative outcome.

"I don't care what you tell me," I said. "I'll sign anything to go to space."

Yes, I understood that severe bodily injury and death were possibilities. It was not my first time around a consent form. One of my job duties is obtaining consents from all the new patient families and explaining to them the risks associated with the various procedures and medications.

Yes, I understood the risks. Yes, I accepted them. I signed the forms.

On the previous visit, I had met with SpaceX's main flight surgeon, Anil, and provided him with my medical history. He and three other physicians had converged around me to do a basic physical exam, listening to my heart and lungs, looking in my ears and feeling the lymph nodes of my neck, and walking me through a neurological exam.

I had been asked to flap my hands back and forth, and I had done so with haste.

One of the physicians laughed. "Wow, you're fast."

I said, "Gotta go to space!"

It turned out to be the most extensive medical evaluation of my life, even more than my checkups at St. Jude. Every system of my body was thoroughly evaluated.

Then some of my tests came back abnormal and more tests were added on. Hearing the word "abnormal" made me want to cry. I felt the most nervous I'd been since getting the call. I'm not a fan of going to the doctor in general, and usually I have Mom next to me, assuring me that everything will be fine. Since I've become a PA, I'm even more nervous when I have to be in the patient role. Too much knowledge can be a bad thing.

I was frightened, but I didn't want to look weak in front of the SpaceX physicians. Astronauts have to be strong, and I needed them to give me the green light.

I endured test after test, until I was finally medically cleared for spaceflight.

My first win.

CHAPTER 11

Inspirati④n

It was on my second trip to SpaceX, on February 1, that Jared announced to the world the first all-civilian mission to space. At that point, only two of four crew members had been selected—Jared and me—but my involvement was still a secret.

Ultimately, the four Inspiration4 crew members would each represent a different pillar of inspiration. Jared would represent "Leadership." As the St. Jude ambassador, I was representing "Hope." The other two seats would be won through contests.

"Generosity" would be claimed by someone who had donated to St. Jude; anyone could donate and be entered to win in a lottery. (My friend Gabrielle would wind up entering more times than I could count.)

"Prosperity" would be won through a contest: Entrepreneurs whose businesses ran through Jared's website Shift4Shop could enter by making a Twitter video about their story and business. That winner would be chosen by an independent panel of judges.

Jared and I had no say in who would be selected as our fellow crew members. I thought constantly about who would be sitting in those seats, hoping they would be cool, hoping we would all get along as well as Jared and I did.

In this first big announcement, Jared revealed that he was leading the mission as a fundraiser for St. Jude. He said that the "Hope" seat had been filled, though he never said who would be sitting in it. He did drop a lot of hints, revealing that it would be a former St. Jude patient currently working as a medical professional for the hospital. He used the pronoun "she."

I wasn't able to tell my colleagues that I was going to space yet. I was nervous about telling them, because I knew the training and mission would take me away from my PA role and consequently they would have to take on a bigger workload while I was away. My colleagues are intelligent people who pay attention. They put the clues together: my mysterious work with fundraising, my conspicuous absence from work on the day of the announcement . . .

Within hours, my co-workers were texting me, telling me how happy they were for me. They didn't ask if I was the one going; they just knew. One said that she had already bought Inspiration4 gear in support of our mission. I neither confirmed nor denied my involvement, but I was so touched by their unwavering support.

The day after the announcement, I wrote in my journal, "People keep saying my life will change. I don't know how much I believe that. I'm just a girl going on a trip. I'll never feel deserving of this opportunity. God has special plans for me that I'll never feel I deserve. Maybe my life will change? Time will tell. But tonight, I feel gratitude, hope, and confidence in this mission."

A few weeks before my announcement as the second crew member, Jared gave me the go-ahead to tell a few close friends my news. Lauren was coming to Memphis for the weekend, and I mentioned that I had some big news to share.

Along with being my cousin, Lauren is my very favorite person and best friend. She always has been. Now she was standing in my guest bedroom doing her makeup in the mirror. I looked at her applying mascara, pursing her mouth and raising her eyebrows as she waved the wand near her face.

"Well," she said, "are you going to tell me your news?"

"Yeah," I said. "It's kind of crazy. I'm going to outer space."

She yelped and spun toward me, nearly painting a stripe down her cheek in the process.

"I could cry," she said as I explained more about the mission. "I'm proud of you. I'm nervous. But I'm just so proud."

I told my other closest friends that I had a big secret, and we made a FaceTime date.

I started the conversation by asking if they had any guesses.

"Dating show?" said one.

"Nope," I said. I hated to disappoint, but no, I was not going to be on *The Bachelor*.

Finally, I told them the truth.

"Wait, what?!"

"No way."

"You've got to be kidding."

They were absolutely stunned when I told them I was going to be an astronaut.

Gabrielle texted the next day. "I still can't believe you're going to fucking space."

Yeah, me either.

The second big announcement about the mission, on February 22, was to reveal my involvement. My mom had come with me on this trip, enjoying a tour of SpaceX while I was in meetings to discuss the entire mission profile. Seeing the facilities and meeting the engineers gave her a lot more confidence in the mission. The plan was for us to fly together from SpaceX to the St. Jude studio in Memphis, where we would reveal to the world that I was the second crew member. We still had no idea who the people were who might be joining us.

Memphis had a freak ice storm at the same time, which froze the underprepared pipes and disrupted the water supply. Instead of flying to Memphis, we flew to New York City to do the announcement.

From a laptop in Jared's kitchen with the skyline of Midtown Manhattan in the background, I told the world I was going to space. I had too many follow-up interviews to count.

At the end of the day, I had hundreds and hundreds of unread texts and messages. It felt like everyone I'd ever known reached out that day to express their shock and excitement for me. I just felt so happy.

Before the announcement, especially when I was going through the extensive medical evaluations, I was so worried that the opportunity would somehow slip away and no one would have ever known. Now the world knew. I didn't have to keep the secret to myself anymore.

I was going to be an astronaut.

In every interview I gave about going to space, I was asked what I was looking forward to the most. Every time, I said that it was the call with the St. Jude patients from space.

By early March, we knew who our two other crew members were going to be. Everyone had gotten there through such different means. Chris Sembroski would take the "Generosity" seat, having gotten his ticket through the lottery, winning out over 72,000 other people. Even more incredible was that it was actually one of his friends who'd won the seat on Inspiration4. That guy couldn't go, and so he gave his ticket to Chris. Dr. Sian Proctor was a geoscientist and a community college professor who had set up her space-themed art and poetry website, Space2Inspire, through

Jared's company. She won the "Prosperity" seat through the viral Twitter video contest. What a prize!

We were definitely the lucky ones.

In late March, we all converged on Cape Canaveral. I liked my other crew members immediately. From the moment we met, Sian exuded the kind of confidence and warmth I would come to see was her signature.

Chris hugged me when we were introduced. "I'm so excited!" he said, and then he hugged me again.

I was able to bring a plus-one to our full-crew announcement in late March at Cape Canaveral. "Hayden," I said, "will you come with me?" I knew he would appreciate the history of Cape Canaveral and geek out on the tour.

"YES!" he said.

It was good to have my brother by my side as we toured the launchpad.

"Do you see that?" he said, pointing at a sprinkler. "I modeled that in college."

I felt so proud of him.

"My brother modeled that!" I kept saying.

We toured the Falcon 9 rocket factory and I made mental notes as the team pointed out the different stages of the rocket and how it would all work together to take us to orbit.

Hayden asked the engineers technical question after technical question:

"How does the thrust vectoring work?"

"What type of propellant do the engines use?"

"How many engines can you lose and still make it to orbit?"

Normally I would have told him to stop, but I knew he was getting all the information he needed to feel this rocket was safe enough for his big sister.

In the days prior to the full crew announcement, we were filmed and interviewed together. I was so excited that we were a diverse crew. Half the crew was female, Sian is Black, and I would be the first person in space with a prosthetic body part, as well as the first pediatric cancer survivor and the youngest American.

I loved hearing Sian's and Chris's backstories and how they each had a lifelong passion for space. Sian had a lot of experience. As an analog astronaut, she had explored remote places on Earth to test conditions for NASA. She'd also been through the NASA astronaut selection process—she made it all the way to the final phone call. That time, she didn't get to go. Now, finally, she had her second once-in-a-lifetime chance.

She smiled as she told me her story. I just wanted to be her best friend.

I was also getting to know Chris, former Space Camp counselor, former air force, current engineer and dad. I learned that he had two daughters, whom he clearly loved so much. I also learned that he possessed no shortage of dad jokes. His attitude and demeanor reminded me of my dad. Dad loved being a dad. He would always engage with my friends, and he really put his family first. I could see a lot of that in Chris.

Then there was Jared. Over the past months, our relationship had taken on a brother/sister vibe. I loved that he could make me crack up, and he also had a thoughtful side to him. I already revered him.

When I travel, it's the relational aspect that really calls to me. What I love about a place is the people. These were going to be my people for the most incredible adventure I'd ever been on.

Immediately following the announcement, we flew from Cape Canaveral to Pennsylvania to start training in something called "the centrifuge." On the plane, my new crewmates and I were handed a thousand posters to sign for donors who had supported our fundraising mission. With each slash of my pen, I wondered why anyone would care about my autograph.

We chatted while we signed. I felt so thankful for these good people and impressed by how well we had all been jibing and interviewing together, despite having only just met in person. I was struck by the feeling that I'd known each of them for years, even though we were just getting started.

We were never given a full rundown of what to expect with training. SpaceX had never had a civilian astronaut crew, so they were building the training as we were going through it and adding additional training here or there as needed. Classic astronaut training includes centrifuge training to become accustomed to the G-forces we would experience with launch and reentry; hypoxia training to learn what our symptoms would be if our cabin's oxygen levels dropped; zero-gravity training to experience floating in microgravity before going to space; and water survival training for when we splashed back down in the ocean. (This sounded the least fun to me.)

I knew we would be studying the ins and outs of our spacecraft, the orbital mechanics, and what we would be doing in space. There would also be several weeks spent in Hawthorne, California, at SpaceX for in-person training in the classroom and in the Dragon capsule simulator. The goal was to practice what it would be like if everything was going right, and also run through what we would do in situations if things went wrong.

I was keeping an open mind and going with the flow. This was the most out of control I had ever felt, but I was at peace knowing we were in good hands and that we would be well trained before we blasted off.

Jared kept reminding us that we had to prove ourselves in order to be signed off by SpaceX to fly on the mission. Being at SpaceX and doing the training didn't mean we were going. Up to that point, being chosen just meant that we got to try.

If we wanted to actually get to space, we would have to earn it. First up, the centrifuge.

CHAPTER 12

G-Monster

The summer I was twenty-one, I did lab research for St. Jude as part of the Pediatric Oncology Education program. Part of the work was analyzing DNA, and it was my job to extract the DNA by running it through a centrifuge.

A centrifuge works by spinning whatever you put in it so intensely that centrifugal force begins to act upon it. Since the DNA in our cells has a lighter molecular weight than the other cellular stuff, like proteins, when the centrifuge spins around, the lighter stuff and the heavier stuff separate. It's like a blender that works in reverse, as if you could take a milkshake and turn it back into Oreos and vanilla ice cream.

Just think about how powerful your blender is. The centrifuge is that and more. So powerful that over the course of that summer, every time I put a vial into the centrifuge, I stepped back and away from the concentrated force of that whirl.

Now I was being asked to get inside of one.

When I said yes to going to space, I was also saying yes to different types of training designed to make us as uncomfortable as possible in order to help us adapt to the shifts in gravitational force that would be a major part of launch and reentry.

Our first crew exercise: the centrifuge.

The moment I first heard about the centrifuge, before we even got to Cape Canaveral for the big crew announcement, I flashed back to being fifteen and the buckle in my leg when my prosthesis gave way. The two years it took to heal were excruciating, and while I'll need a new prosthetic someday, it's not something I like to think about or talk about.

Cancer is something I had. I got better. I don't have cancer anymore. What I do still have, what I'll always have, is a rod in my leg that I worry could break at any moment. And now I was going to take my leg into highly stressed conditions, on purpose.

Consider that Earth gravity has a G-force of one. You can't feel it, but it's strong enough to keep you from floating away. The centrifuge was going to be my first introduction to six Gs, or six times the normal amount of gravity.

If my rod could break when I simply stood up, if I was already worried about slipping and hurting myself during a light drizzle, what could the centrifuge do to me?

I consulted with Kidd. "Do you know of anyone else who's done centrifuge training with a prosthetic body part?"

"Nope."

"Cool."

On the outside, I was all confidence. Of course I could do it! No problem! Bring on the G-forces!

What I wanted was reassurance. If only someone could tell me my leg would be okay.

I was at home about a week before we started training, and I went to visit Dr. Doom, who had reached out to say that he had a gift for me.

I hadn't seen Dr. Doom as a patient for about ten years, even though a yearly checkup was recommended. As a cancer survivor, I do not like going to the doctor and putting on that gown. Yes, I'm a medical professional and I work in a hospital every day. Yes, I love working with my kids and my families. It's just my own appointments I dread. They give me the sweats. I have a pretty complicated history with visiting the doctor. Ask anyone who has received a terrifying diagnosis how they feel about a casual doctor's visit. I am pretty sure I'm not alone in this.

Despite my decade-long reluctance to stop by as a patient, it was really nice to see him. He presented me with an official St. Jude orthopedics sweatshirt.

Then I asked him the question that had been burning up my mind. "Do you think it's going to be okay for me to do centrifuge training? Can my leg handle all of those G-forces?" I trusted him to be real with me: Even when I was a kid, while everyone else was trying to shield me from reality, he would tell me the worst things that could happen.

"Probably," Dr. Doom said.

I breathed a sigh of relief. A "probably" from him is like a big smile plus a high five from somebody else.

Then he said, "But . . ."

. . . and he picked up one of the prosthesis models he had on his shelf.

"Your prosthesis can bend this way," said Dr. Doom, giving me a visual demonstration of the normal movements of the prosthesis. "And it can bend this way."

I nodded.

"The problem is if it goes like this." And he separated the prosthesis in two, holding the pieces apart widely.

I thought I was going to throw up. This was my absolute worst nightmare. "That can't happen," I said. "If that happens, I will die."

His nurse practitioner, whom I've known for years, is a woman I really respect. "Dr. Neel," she said, "that isn't going to happen. She has bone holding it in, and muscle, and skin."

She turned to me. "Hayley, your leg is *not* going to split in two."

At least one of us was sure.

We got to the hotel in Pennsylvania the evening before centrifuge training, and I fell asleep quickly, exhausted from the last few days. As I was drifting off to sleep, I thought about what the next day would bring. Hayley versus the centrifuge. This would be my first real activity with my new team, our first astronaut training. This was where we would show one another what we were made of, and I was determined to show them how tough I was.

My butterflies flew away in the morning light, replaced by adrenaline and excitement as I stepped into my brand-new flight suit. I had never worn anything like it. The material was thick and hadn't been broken in yet, and it was covered in zippers. I kept discovering pockets I hadn't noticed before.

I looked in the mirror. I looked so badass and official. This unbelievable getup was the most amazing thing I had ever seen myself in. I am a person who loves a costume; my friends and I have been known to wear tutus and fun Mardi Gras outfits. But this was something else. Not pretend. Not dress-up. All real. There's nothing like a flight suit to remind you that you've got what it takes. The confidence boost was through the roof.

Our energy in the car on the drive over to the building that housed the centrifuge was really loud and all over the place. We were all talking and laughing, riding this wave of excitement. We had each traveled such a different path to this moment, and I was looking forward to getting to know them over this journey.

"I want to talk with you guys about something," Jared said, cutting through the electric energy in the car with his firmest "I'm serious" tone. One thing about Jared is that he can be very funny, but when he is serious, he means it.

"You've all been selected, but we still have to *earn* our spots," he reminded us again.

This was some kind of pep talk. Not exactly the "You can do it!" kind. More like "We need to do this."

"SpaceX will not fly us if we're not ready," he said, making it super clear what the stakes were.

We all got really quiet. I took a breath and told myself to focus. Jared was a fighter jet pilot. Sian was a geoscientist and private pilot. Chris had been in the air force. And then there was me. I was the youngest and the smallest. I had medical training but no flight experience, unless you counted commercial window seats. I was the one who had the prosthesis—but I didn't want any special treatment. I wasn't here to be a passenger. I was the medical officer.

I belonged here. I was bringing my own experience, and they needed me.

I knew it. I just needed them to know it.

We arrived at the building housing the centrifuge. It was contained in a massive room that had large windows for maximum viewing pleasure—the pleasure of the viewer, not of the person being centrifuged, obviously. The machine itself was basically a little capsule attached to a motor that would spin it all around the room. They gave us a quick tour, and then we took a deep dive into the physics of G-forces and the profiles of launch and reentry.

We would be going into the centrifuge one by one. We wouldn't be interacting after our experiences until we each had had a turn, they explained, because they didn't want us influencing one another.

They gave us the lineup: first Jared, then Chris, then Sian, and finally me.

Jared climbed into the capsule. For the rest of us, there was nothing to do but watch. I looked through the windows as the small capsule he was in began to hurl around, whipping through the space at an alarming speed. There was also a monitor set up that allowed us to watch what was happening inside the centrifuge.

I could see Jared's face contorting under the power of this strange new pressure. When he hit high G-forces, I saw how hard he was concentrating on breathing. We had been instructed in techniques that would keep our lungs inflated despite all that pressure on our chests.

SpaceX had designed the profiles of that machine to feel like launch and reentry by taking previous readings from other spacecrafts and using their G-force data as input in our centrifuge. The whole point was for us to feel launch- and reentry-related G-forces and adapt to them. They wanted to see if we could handle it psychologically as well as physically, because it's an experience very much out of one's control, uncomfortable, and claustrophobic.

I was not too thrilled to be going last. Mostly because I hate waiting.

Next up was Chris. Who threw up. And then gave us a thumbs-up to let us know he was okay. (Classic Chris.)

Sian gave me a reassuring hug before she slipped out the door. (Classic Sian.) Then only I was left, just me and my thoughts.

Classic Hayley: I always pray right before something big. I don't shut my eyes, and I'm not one of those who prays with other people. Instead, my prayers are always very introspective, internal, and personal: "Thank you for my health. Thank you for my family's health. Thanks for my friends and family. Thanks for my dream job. Thanks for my house, for Scarlett. For this opportunity to go to space. And please, please let my leg be okay."

My turn. I walked into the room in my flight suit, channeling that badass feeling as I climbed into the centrifuge. The interior of the capsule looked like you'd imagine a spaceship, with a flight seat, a mock control deck, and a G-force meter.

I sat down and strapped in, thinking about the centrifuge in the St. Jude lab. Now I was the test sample.

My Apple Watch registered my heartbeat, which was really fast. I took a deep breath and looked for my calm. I wasn't there only for me. I was there for something bigger.

Instead of focusing on what I was afraid of—pain, the unknown—I focused on the larger picture and the real reason I was there. *I have to do this,* I told myself. *Not just for me. For everyone coming after me.* This wasn't about this centrifuge or this exercise. It was about going to space. It was about representing people who weren't physically perfect. In order for more people with prosthetics to follow in my footsteps, first I would have to get to space.

I felt like there was a lot riding on my shoulders, but in a beautiful, empowering way.

They turned the centrifuge on and the whole capsule shifted its angle; suddenly I was on my back, in the same position I would be during launch and reentry.

The capsule started to move, and I could feel the G-force vectors going through my chest, front to back, with so much force that it felt like multiple people were lying on top of my chest.

When I first started learning about the concept of G-forces, it felt abstract, like an idea. But let me tell you, G-forces are real. Gravity is like something pressing on you. We don't notice it on Earth, but when the force gets stronger, that's exactly what it feels like.

At first, it was hard to take a breath. The key was not to get anxious about how hard it was to breathe. I spoke to myself, telling myself that I just had to get through it, that I'd be able to take a deep breath in a minute.

I had been so worried that it was going to be painful. But it wasn't. It just felt like—like a really, really tight hug.

Then my vomit bag came loose and whipped past my face and flew around the capsule. And again. And again. They had graciously included a receptacle in case we didn't feel our best in this extreme version of a theme park teacup ride, my favorite Disney ride.

Then the world stopped moving. I stopped moving. The vomit bag stopped moving. I could breathe again. I hadn't vomited. My leg was still in one piece.

I sent up a quick prayer of thanks that it had gone so well.

Before we left the facility, we stood together in front of the centrifuge in our flight suits for a graduation ceremony in which we were all given certificates that heralded the successful completion of our first training exercise.

"I'm looking forward to seeing you all fly," said the instructor who was giving us the certificates. It was very sweet.

Just a short while before, I had been all nerves. Now I was riding high. Just a few weeks back, the others had been strangers. Now I felt very close to them, like we really were a crew. I was part of a family, and the Inspiration4 team was much bigger than us four.

Kidd called me the day after our triumph in the centrifuge.

"Hey, Kidd!" I answered.

"Hey, G-Monster!" he replied.

I started laughing.

"We were watching you in the centrifuge," Kidd said. "You nailed it, G-Monster."

I smiled so big in that moment, I couldn't stop smiling. The Corsentinos called me Haylstorm. My college rapper name was Arce-no-mercy. Now I would add G-Monster to the list of names that made me feel seen and loved. When they give you a nickname, it means you belong.

G-Monster 1, centrifuge 0.

CHAPTER 13

Upper Limits

According to my dating app profile, I enjoy hiking. More truthfully, what I enjoy is the idea of hiking and the views at the end of the hike. I love to be active, but I try not to take my leg into uncertain circumstances (you know, like centrifuges).

Memphis has temperate weather, but it does snow a little, and it rains a lot. My habit is to stay in when either of those things is happening. When it's cold out, my leg has a constant ache. So when it starts to snow, I do not run directly outside like some people might; instead, I grab another blanket and get cozy with Scarlett. So when Jared added a training exercise on Mt. Rainier to our schedule, I shuddered. There's the discomfort that the cold causes, but mostly, I didn't want to slip and fall.

Jared was into the idea of our "getting comfortable being uncomfortable." Mt. Rainier, in Washington State, is an active volcano that rises more than fourteen thousand feet to a glaciated peak. This mountain is so big and icy that it feeds five rivers. And we were going to climb it, which meant we would also have to descend it. It wasn't meant to be easy. It was meant to serve as crew bonding. If the goal was discomfort, Jared's idea was already working.

I knew that if I wanted to go to space, I couldn't pick and choose which training activities I wanted to do and which I did not. This was not a training buffet. I couldn't offer to get back in the centrifuge and skip the climb. If I wanted to be on this team, I needed to prove to Jared, Chris, and Sian—and to myself—that I could do it, training exercise by training exercise.

I reminded myself that hiking is a necessary part of travel and that I have put my leg through some treks over the years in order to get the best views. I didn't do the weeklong hike through the wilderness to get to Machu Picchu; instead, I took the train with the glass ceiling. But the day we went, we did hike up Machu Picchu mountain to get the most scenic outlook on the Wonder of the World. The view from above tree line was worth being sore the whole week after.

I have a gift for optimism, so I leaned into that.

I wasn't a gym person before I trained for space. After my diagnosis, I hadn't been able to take PE classes, so aside from an occasional yoga class as an adult, exercise hadn't become part of my routine. I was told that greater muscle tone would be beneficial to operating in microgravity and would also help my tolerance of the high G-forces during launch and reentry.

When I admitted to my flight surgeon how little exercise I was getting, he connected me with a trainer. I was feeling motivated to get moving! In the first workout she sent me, I walked on the treadmill for twelve minutes and felt absolutely miserable.

She sent me workouts every day, motivated me, and held me accountable. I loved seeing my progress as I could lift heavier weights, hit higher durations of time on the treadmill and exercise bike, and even balance better on one leg. My friends noticed my arms were getting more toned (which I loved to hear), and overall I felt stronger.

Every day, I increased my treadmill time and incline until I was walking for over an hour with the incline as high as I could get it. A few days before the climb up Mt. Rainier, I hit seventy minutes on the treadmill with ease. I was so proud of how far I had come in that month of training.

I called Jared and said, "I hit seventy minutes on the treadmill. Do you think I'm ready for this?"

His "yes" in response wasn't quite as emphatic as I had hoped.

I was worried about a fall, but I still felt the rush of optimism surging through me.

Training in space happened in blocks of time. When I wasn't training, I was back at St. Jude working with my patients. It was an intense schedule.

One day, a woman stopped me in the hallway. She was with a little girl who was a patient.

"Are you Hayley?" the mom said. "The girl who's going to space?"

"Yes," I said, smiling at them. "That's me!"

I was used to people asking me about my space journey now, and I loved to share about it. I don't know whom it made happier, me or my patients.

What I wasn't expecting was that this woman would start to cry.

"My daughter has been having a hard time," she told me. "Last night all she did was cry, and the only thing I could think to do was show her your story. Thank you for doing this mission and for giving her hope."

I turned to the little girl with the smooth head and bright eyes and invited her to join me in the closest available seating, a cluster of chairs off to the side where we could speak privately.

"What's been going on?"

It turned out that her siblings had been running and jumping the night before, and she felt sad that she couldn't do that with them.

"I can't run or jump like my little brother either," I said. "But it's not going to stop me from going to space."

She looked at me, processing this.

"You don't need to run or jump to do big things," I told her.

I don't know how impactful that moment was for her, but I can tell you that it was transformative for me. I had been so focused on raising money to help these kids. That little girl reminded me that we were giving the kids something more. We were giving them hope.

After a week at work, it was back to training: climbing Mt. Rainier. We met our guides at base camp the day before the hike began, and they ran through the plan and gave us a safety briefing. Then we showed them our gear. I had gone to an outdoorsy store in Memphis and, with the help of several sales associates, picked up the gear I would need. While I was waiting for our mountain guides to approve my gear and make any suggestions, I was taking off price tags and shoving them into my new backpack that was half my size.

Kidd showed me how to put crampons onto the mountaineering shoes I'd rented and walked me through how to use my new gear. He had years of experience, so I took to heart all the advice he gave me.

"Is your backpack too heavy?" he asked.

"No," I replied, though I wasn't sure how true that was.

The small town outside of Seattle was blanketed in fog. I couldn't even see the supposed mountain we would be climbing the next day.

We were bunking at the main lodge in town for the night. After unpacking and repacking my backpack, I stretched out in the bed, savoring every minute I had left in the warmth, knowing we would be leaving early for the climb. Bliss.

Followed shortly by the buzz of my alarm at 6:00 A.M.

Early the next morning, sitting in the back seat of the SUV, listening closely as Jared provided us with last-minute tips for the climb, I thought, *I'm ready . . . to get this over with.*

We convened at the trailhead with our guides, SpaceX physician Jo, Kidd, and our other mission director, Leif. Leif was also a friend of Jared's, a former air force test pilot, and an aerospace engineer. A film crew for our documentary was there to capture the glamour of the climb.

They positioned us in a single-file line, and I took my place behind Sian. I asked Jared to be behind me because I felt like he had the best potential to catch me if I fell. We grabbed our poles and set off. Our guides instructed us to take one step at a time, leaning on the leg that was planted with each step and taking sharp exhales. The guides were all males except for a woman around my age.

I watched my feet for the next nine and a half hours, cautiously taking step after step. The strategy, I learned, is to step into the flattened snow where the person in front of you has already stepped, because it's sturdier. Still, there were moments where I would step forward, the snow beneath my foot would give way, and I would stumble. When we went up steep inclines, I sometimes had trouble reaching the next step. When we went downhill, I sometimes had trouble bending my leg enough to take a large step down.

My crew was constantly checking in on me.

"How's the leg holding up?" was a constant refrain.

"Fine," I would reply, or "Good." I had taken a large dose of ibuprofen prior to starting up the mountain, and so far, it was keeping the pain in check.

I was not planning to say otherwise until I was really struggling.

Every forty-five to sixty minutes, we stopped for a break, where we were encouraged to reapply sunscreen, eat snacks, and drink water. I would sit on my backpack in the snow with Leif and soak in his natural optimism. Since we were both fans of dark chocolate, we shared our respective stashes. I took selfies with my crew, the amount of ice frozen to my hair in the photos increasing at each stop.

I asked the guides for updates on how far we had already climbed, in percentages. Around the 65 percent mark, I started to feel woozy. Not just woozy. I saw dark stars creeping into my peripheral vision. I knew what that meant.

I mentioned to the physician Jo that I was feeling like I was getting close to fainting, just so she would be aware. My tone was as casual as possible. I felt like I could power through it though, and I took large gulps of water.

I started reciting a new mantra to myself: *Can't faint, must go to space, can't faint, must go to space.*

The whole way up was difficult for me. There wasn't one easy moment. I was very aware of the camera crew documenting every step. A few times when I had trouble taking a step up an especially steep incline, I felt the camera's glare watching me struggle. I was a tad annoyed, because I was trying my very hardest and I didn't want to look weak on camera.

Thinking about my patients helped me through. We ask a lot of them, these kids going through cancer treatment, and they don't complain. What they're doing is a lot harder than just climbing up a mountain. Their positive attitudes motivated me to keep on going. I thought about how I was doing that mountain climb, and the whole mission to space, for them. It taught me a lot about how important it is to have inspiration in our lives.

I also thought about my own cancer journey. My mom's mantra throughout my treatment was "One day at a time." We couldn't think about how many more rounds of chemo I had left to go, because if we did, it was too overwhelming. We could only concentrate on that one day, that one moment. I couldn't think about how much longer I had left to climb to reach our destination. I had to concentrate on my one step, followed by one other step.

The other thing that sustained me was my dad's belief in me.

Just after Dad died, we were going through some old bins in our garage when I found the board he had inscribed to me when we got our brown belts: "I'm prouder of you for conquering your fear than for earning your brown belt," he'd written. I took it home and put it in my closet so that I could see it every day.

My dad wouldn't have cared if I made it up that mountain or not. He would have cared only that I'd tried. He would have wanted me to push through my fears so that I could do what I wanted to do, no matter the obstacles.

With his voice in my ear, I kept going.

The guides recommended that I work on sharpening my exhales, since we were reaching higher and higher altitudes. After a few minutes of trying that, I felt the stars in my peripheral vision subside and I felt stronger.

We were in the clouds throughout the climb, so we could never see more than a few yards ahead of us. This helped me stay in the moment. I was forced to concentrate on where we were and not get overwhelmed by how much we had ahead of us.

I asked the guides to tell me stories to keep me distracted, and they told me about other climbs they had taken on much higher mountains for much longer durations. Suddenly this climb didn't seem so bad.

Nine hours in, we were so close but still so far. At this point the clouds were clearing and we could see Camp Muir, our goal, in the distance, sitting at just 10,188 feet in elevation. Each step felt heavier, and each breath was becoming more labored.

"Fifteen minutes and we're there," Jared encouraged us.

"We can do anything for fifteen minutes!" I replied.

I used that same approach with my patients when I worked in the emergency department in Louisiana. Just before I would give them an injection of lidocaine, which stings terribly as it's being administered, I'd say, "Breathe through it; it will last only fifteen seconds, and you can do anything for fifteen seconds."

The four of us changed formation. Now instead of being in a single file, we walked side by side the rest of the way up. A victory climb. We took our final steps into Camp Muir, where Kidd was already waiting to congratulate us. Jared pulled us into a group hug (very uncharacteristic) and told us how proud he was of all of us. He couldn't stop smiling.

"You were the only one talking when we hit the high altitudes," he told me. "You seemed to actually get stronger the longer we hiked."

The whole way up, I was looking forward to the dinner I'd brought with me. The guides had hot water for us, but we had been expected to pack our own food. Always a picky eater, before this trip, I couldn't bear the thought of freeze-dried food, so I'd packed several ham-and-cheese sandwiches along with potato chips for the trip. Rookie mistake. My ham-and-cheese sandwiches were frozen solid, and all of the chips had gotten crushed in my backpack.

I bummed a freeze-dried pasta from Jo, our doctor, and discovered to my relief that I loved it. Everything tastes better at altitude.

We were each assigned a tent where we'd be sleeping for the next two nights. Thankfully, they had already been set up for us. It was such a relief to take off the giant mountaineering shoes and several layers of coats. Despite being in freezing temperatures and sleeping on the ground without a pillow, I slept deeply, curled up in my subzero sleeping bag. Exhaustion is life's natural sleep aid.

The next morning, the clouds had cleared. Sian had brought plenty of tea to share, and I drank it gratefully from a mug while looking out at the spectacular view: stunning mountain peaks all around us, alpine beauty in blue and white as far as the eye could see.

After breakfast, we hiked to a nearby crevasse and took turns being lowered into it by our guides. We didn't have to do it but all opted for it; when else would I have the chance to be lowered into an ice cave?

Now that I was going to space, things that would have scared me in the past were put in a new light. *I can do ANYTHING.* The words echoed through me, offering the promise of so much more.

Later, after our hike to the crevasse, one of the documentary guys offered to let me use his satellite phone to call my mom and tell her I was safe. She wasn't expecting to hear from me, so her first question was "Are you at the hospital?"

"No, I'm not at the hospital," I assured her.

She asked me how my leg was.

"Great! Turns out I love mountaineering," I told her, laughing.

In the evening, we sat around the campsite, taking in the dazzling pinks, oranges, and yellows of the gorgeous sunset set against the backdrop of the mountains casting their long, dark-purple shadows across the wild land. Being on that mountain, looking out at the vastness, made me pause and reflect again about the same thoughts I'd been having for months, but now with a deeper intensity: How lucky I was to have this opportunity to go to space, and with such a great crew. How happy I was to be alive. How proud I was of my leg and how it was holding up.

On our second and last evening in camp, Chris surprised us with a round of Fireball whiskey he had brought, you know, "for warmth." I'd never loved him more.

The next day, it was time to hike down the mountain. What goes up must come down—the beauty of gravity. We packed up and started our trek down. Soon I was way behind the pack. Fun fact about my leg: While other people usually find going down easier, it's much harder for me. The muscles it takes to balance going down stairs or going down mountains are much weaker in my leg.

Jared and the female guide stayed near me to catch me if I fell. Still, I could only take baby steps. One of the guides offered to put me on a sled for the rest of the way. At first, I declined, being stubborn, but after an hour or two of moving at a snail's pace, I took him up on the offer. Guides took each corner to control the sled as we went downward.

I glided past groups on their way up the mountain, waving to them and smiling. The hard part was over for us. When we got back down to the bottom, we each had a celebratory beer.

We had done it! We had trekked to the top and back again! I felt absolutely elated. I hadn't known how long the hike would be before we set off, and I surely did *not* know I would be hiking for nine and a half hours up a mountain. If I had known that ahead of time, I would not have even attempted it. Instead, I had crushed it. There are so many things in life that limit us naturally; the last thing we need to do is put the handcuffs on ourselves.

In that moment I understood that I had been putting limits on myself that did not need to be there.

I arrived back in Memphis at 2:00 A.M. and fell wearily into bed. Just a few hours later, at 7:00 A.M., I was back at work.

That afternoon, I met with my boss's boss, the director of advanced practice providers at St. Jude. I had gotten the meeting invite when we arrived back at base camp the evening before.

"How are you doing?" she asked me when I sat down in her office.

"Ummm, tired," I said, smiling at her. "I just spent the last three days on a mountain."

I had been struggling with balancing training, studying, working out, and media appearances with my full-time position, so I had reached out and asked for some help. If I could organize a day or two off per week as dedicated Inspiration4 time, I thought I could make it all work.

"That's what I wanted to talk to you about," she said. "We're recommending you take leave until after your mission, in about seven and a half months."

I burst into tears. "But this job is my purpose in life," I said.

"Your manager told me this was going to make you sad," she said, handing me a box of tissues.

While I blubbered in her office, she comforted me by telling me I was still helping the kids, just in a different way.

"You'll be able to impact so many kids through Inspiration4," she assured me, "even more than you would on a day-to-day basis."

I thought about what she was saying.

"We want you to enjoy this time. We want you to spend time with your family."

My family? What did my family have to do with this?

The cold hand of realization touched the back of my neck. *They want me to have quality time with my family now. Because I could die.*

The thought echoed in my head: *What if I die? What if I've spent my entire life working toward getting a job at St. Jude and this turns out to be my last week at my dream job?*

The emotions overtook me. Until that moment I really hadn't thought much about dying.

Now I was really crying.

"This is exhibit A that I'm exhausted," I explained through my sobs. "I don't normally cry like this."

I knew that I couldn't keep up with what I was doing. I knew that being put on leave was the best thing for me. I couldn't argue with the fact that training was just starting to increase in pace and intensity, and I was already struggling.

Stepping away from my patients felt like the ultimate sacrifice, one I did not want to make. It hurt, but I eventually stopped crying and agreed that a temporary step away from my position to focus on becoming an astronaut was the right move. Emphasis on "temporary."

After rounds on my last day, my attending physician asked if I would miss my PA job.

"Yes," I said, and turned away so that he wouldn't see my eyes fill with tears.

I told myself that I'd still be very connected to St. Jude, and that I'd be able to talk to the kids from space. I told myself that I'd be back. Even though I was walking toward something new and exciting, I felt deeply that I was walking away from the only thing I had ever wanted.

Just because I love new experiences doesn't mean that change—even temporary change—is easy.

CHAPTER 14

Inside the Simulations

I threw myself full force into training, the majority of which was spent in the Dragon capsule simulator at the astronaut training center in California. At times, we'd be in the simulator for hours on end, getting to know the model of our living quarters, working through unexpected situations, prepping for whatever emergency they would throw at us next. It was never leisurely, always busy.

We wore our training space suits only when the situation demanded it. The rest of the time, we'd be hanging out in casual Inspiration4 gear and fun colorful socks, since you don't need shoes in space. On the actual mission, we would be wearing our space suits for launch, for reentry, and if there were any emergency situations.

Midway through training, we practiced our first launch simulation, as if it were launch day, with mission control present, as well as the crew who would be loading us into the capsule on launch day. Jared reminded us that we needed to take it seriously and focus. With his words ringing in my ears, I felt slightly nervous and reminded myself I had every reason to feel confident. I had pushed through my fears and perceived limitations again and again over the previous weeks and months. If training was meant to get me comfortable with being uncomfortable, it was working. I had made it through the centrifuge, and I had made it up a mountain. I was powerful. I felt strong and I felt capable, ready for anything.

On that particular day, we were training in real time. At launch, SpaceX astronauts are strapped into the capsule about two and a half hours before liftoff. The majority of those two and a half hours is spent going through launch procedures, but for part of that time, our main job would be to wait while the ground crew took care of everything that needed to be done in order for us to launch safely.

Today part of what we would be practicing was waiting. No sweat.

We sat there and sat there. And sat there. After how mentally and physically taxing other parts of training had been, the sitting part didn't seem so bad. What I hadn't anticipated was how sitting in that one position for so long would affect my leg. As we sat and sat, I could feel my leg start to stiffen. Because of the angle of the seat, it was forced into a bent position and I couldn't move it.

I took a deep breath and told myself it would pass.

Stiffness became pain. The pain intensified.

I looked around. Sitting there didn't seem to be bothering anyone else.

I started to feel a lot of pain. I kept telling myself that it would go away. But the aching grew worse. I could feel my face starting to grimace.

I was on audio with my crew. They were cracking jokes here and there, but I was silent.

Finally, I said, "Y'all, don't say anything to SpaceX, but I'm in a lot of pain."

Jared, Chris, and Sian all piped up: "Are you sure? Anything we can do?"

"Yeah, don't say anything."

I could feel my face really grimacing now, my brows knit tight. The ache was so intense that I was sweating.

I took a breath and said, "Don't tell SpaceX, but I really am hurting."

Jared immediately called ground. "Hey, we need to stop the simulation. One of our crew members is having trouble."

My commander knew what I needed, even when I had trouble asking for it. And then they shut it all down, they opened the doors, and a crowd of people ran in yelling, "What's going on? What's going on?"

I hate attention with medical problems, because I've had enough of that. I can't handle all the drama. I wished I hadn't said anything, I wished I could rewind time or just make everyone calm down. And so I said, "Everybody's being really dramatic."

Someone said, "Get the flight surgeon."

The flight surgeon, my friend Anil, came in.

"My leg's killing me," I told him. "I'm so sorry."

And I burst into tears. I cried and cried.

"My leg did *great* climbing Mt. Rainier, and it's being taken out by sitting in a chair?" I said through my tears. I was so embarrassed that the simulation had to be paused on my account. I couldn't believe it. Painful feelings began to resurface—in that moment, I felt like the cancer girl once again, like I couldn't keep up.

I kept saying, "I'm still the toughest crew member!"

Anil said, "Look. No one's ever done this before with a prosthesis. You're the first one. This is new for us, and we've got to figure this out. We will fix this."

That made me feel hopeful.

He was right. We needed to figure out a seat that would work for me, and many more people to come.

When I could finally focus with that mindset of helping other people who would come after me, I stopped feeling defeated and started to feel comforted. I knew that people with prosthetic body parts would one day be going to space after me, and I knew that figuring this out for me was a big step in making their spaceflight more possible. I was making space more accessible for people like me, who aren't physically perfect.

I had pain that entire day into the next day, because I had waited too long to speak my truth and ask for help.

SpaceX took the problem very seriously. They had a "We've got to fix this" attitude. They held meetings the very next day to redesign my footrest, and they made a new seat for me where my leg was more extended so that I had no pain on launch day.

Speaking up was another kind of pushing through the fear, another pathway to conquering what scares me. Note to self: Always speak up when you are in pain. You should never have to feel bad about the experience you're having in your body.

The next challenge thrown our way was altitude chamber training. We flew to Durham, North Carolina, to visit the Duke Hyperbaric Medicine Clinic so that we could be exposed to two environments: high carbon dioxide (CO_2) and low oxygen (O_2).

Since everyone presents with different symptoms when carbon dioxide and oxygen levels go up or down, the goal was for each of us to experience and understand how we would react to those shifts. This way, if there was a problem with our environment in space and the levels were off, we'd be able to detect it quickly based on our symptoms. We started with a classroom lesson, learning all about the physiology behind hypercapnia (high carbon dioxide) and hypoxia (low oxygen).

Hypercapnia training required us only to breathe into a fancy tube that would basically recycle our own breath, increasing our CO_2. I felt tingly and light-headed, which was followed by a killer headache.

For hypoxia training, we were placed in an ancient-looking chamber with creepy gas masks hanging from the ceiling. With our oxygen masks on, we were brought to simulated high altitude with a resulting lack of oxygen. Since hypoxia can cause mental slowness and confusion, the instructors had a list of tasks for us that were designed to demonstrate our working abilities in that setting.

Chris went first. He made some comments that had me questioning if it was regular goofy Chris or if it was the hypoxia talking.

Then it was my turn to take my mask off. I had an O_2 monitor on my finger to track my dropping oxygen levels. A timer was set for two minutes, and I had a list of tasks to complete. If my symptoms were severe, they would tag me out early.

I initially felt fine, then became more self-aware. I was feeling the effects of the hypoxia. I just felt so happy. That was my main symptom: euphoria, which, as it turns out, is fairly uncommon, so don't try this at home.

I turned to Chris with a smile and said, "I feel *great*!"

They asked me to count backward from one thousand and write the numbers on a piece of paper. I started fast because I wanted to see how far I could make it (and make it further than my crewmates).

As time went on and I became more hypoxic, my handwriting became sloppy. I made it to 900, then stopped in my tracks. What's one number less than 900? I couldn't think of it. I wrote 929, then scratched it out. Again I wrote 929 and scratched it out. I looked at Chris. Then it came to me: 899! I wrote out a few more and looked at my O_2 saturation on the monitor. It was 38 percent. For the record, normal is 95–100 percent.

Normally this would have really freaked me out, but I felt calm. I was having fun and I trusted the team running the training. A few seconds after I put my oxygen mask back on, the euphoria dissipated, and I felt normal. I looked down at my worksheet and saw the increasingly messy handwriting and laughed.

No doubt, I'd be able to recognize that hypoxic feeling again.

Our reward for climbing inside all of these simulations: We went to Nevada to take a zero-gravity flight so we could experience microgravity before we got to space.

In order to achieve zero gravity, the plane flies through multiple parabolas. Imagine a curve like a half circle with more of a dip, like the shape of a roller coaster. That's the path the plane takes. It starts by flying at cruising altitude; then the nose is tipped upward, and it begins to ascend straight up.

Please note that this flight has historically been nicknamed the "vomit comet."

On the uphill, lying flat on our backs, we felt slightly increased G-forces but nothing compared to centrifuge training.

I reminded myself that I would be okay no matter what happened, whispering my new mantra to myself: "There are a lot of things you can't control. And that's okay."

I could handle anything the year threw at me, either in my control or out of my control, if it meant I got to go to space. Whatever the experience was like, I would learn something, just as I had in the capsule. Everything was going to work out.

As we reached the top of the curve, the tip of the plane pointed downward, and on the way down, for twenty to thirty seconds we experienced weightlessness.

We did this sixteen times, which equals sixteen parabolas of pure chaos. Bodies flying everywhere, running into one another, people scream-laughing. I got kicked in the face. A few times I wasn't sure which way was up or down. Then gravity would suddenly come back, and we would hit the deck pretty hard.

Navigating microgravity is harder than astronauts make it look! I hit the walls and ceiling several times before I learned that the harder you hit a surface, the harder it will propel you in the opposite direction. The best way to change directions if you're heading toward a wall is to lightly tap it. It was like my mantra:

There were things I could not control, and I didn't need to try. I just needed to go with whatever was happening.

The result was that my movements became somewhat more controlled with each parabola, and I was able to have more and more fun as I explored what weightlessness had to offer. I went into it thinking it would be pretty cool, but I had no idea how much fun I would have. I even did flips.

I could barely contain my excitement knowing I had three whole days of nonstop zero gravity coming up.

CHAPTER 15

Nova

I met the Dragon for the first time in July.

The crew and our SpaceX training team were down in Cape Canaveral for a few days of training, walking through what launch day would look like, as well as running through launchpad emergency situations.

And there it was, finally, our Dragon capsule. It had already been to space once, taking a crew of NASA astronauts to the International Space Station. It was early in the refurbishment process, but it was so exciting to look at it and think, *That is our spacecraft. That will be our home in space for three days!*

Our Falcon 9 rocket booster was there too. The first stage of the rocket is reusable, and this would be its third mission. I had seen a few boosters before, and every time I did, I was taken aback by their sheer height. Ours was covered in soot from its previous missions, and we signed our names in the soot. Of course, I then wiped the space dust on my trainer.

There was a lot of ground to cover, and one very important part of it was the food.

My friends would always ask me what I was going to eat in space. I told them that I didn't really care; I could eat anything for three days if it meant I was in space. When I first tasted some of the food options they were offering us, I wasn't as sure.

On our spacecraft we couldn't heat food, and we couldn't rehydrate freeze-dried food, so on food-tasting day we were presented with a variety of shelf-stable options. The main concern for the engineers who planned our meals was to choose food that wouldn't create crumbs. Crumbs could become foreign object debris (FOD), which could float through the air and get in someone's eyes or even affect hardware. For example, instead of bread, we were given tortillas as an option because they produced fewer crumbs.

I was optimistic when I saw the pile of different types of food encased in various wrappers and packaging, but after a few tries of shelf-stable pasta and meats, my smile faded. I'd been spending my adult life working to grow out of my picky-eater habits. Between my love of travel and love of trendy restaurants, I'd had to become open to trying foods that I couldn't always pronounce or that I didn't really want to know what was in them.

There were also "food cubes" that the food engineer was very proud of. They were small, brightly colored cubes that were made of salmon, avocado, and kale. Luckily for us, we were the first crew he had presented them to. They were our contingency food, if we were in space longer than we planned or had to land at an unsupported location and ran out of regular food, since these were full of nutrients. He also told us that we could choose them for our regular meals or for snacks, if we wanted. I tried to keep an open mind as I took a bite, but I immediately gagged. The texture was gelatinous but somehow also chalky. And the taste was . . . as you would imagine shelf-stable salmon mixed with kale would taste. "I don't think I'm going to eat much in space," I told Sian as I wiped my eyes. My crew shared my viewpoint.

Jared tried some colors of the cubes but refused to try the others. The food engineer was visibly offended.

"They're not that bad, guys! I ate these for lunch the other day."

"Where's the astronaut ice cream?" I asked, assuming we would have the freeze-dried dessert that every kid who's ever visited NASA has enjoyed. I was told in response that "astronaut ice cream" is not actually eaten in space, as it creates too much FOD. So yeah, our childhood was a lie.

None of us found many options we were happy with, so the engineers took our feedback and later presented us with another food-tasting experience. This time they served wine. We wouldn't have alcohol in space, but they really wanted us to like the second food tasting. The choices for round two were less healthy and more in the comfort foods category. At the end of the day, I selected white pizza with bacon and jalapeños, bagels with cream cheese, bacon, shelf-stable cheese, salami sticks, almonds, tortillas, peanut butter, Skittles, and M&M's. Plenty of options for three days.

Even though I wasn't working as a PA at St. Jude, I was often back on campus. One of my big responsibilities as the St. Jude ambassador was to do media appearances to raise awareness of our Inspiration4 mission and the mission of St. Jude. I was very aware of how important this was. No matter what else was going on with my training, I always tried to show up for interviews with all of my energy and joy, because I felt personally responsible for achieving the $200 million fundraising goal.

In the months before I went to space, I did hundreds of interviews from the studio at St. Jude. I appeared on every major U.S. news network, on international networks, in print, and on social media channels.

One morning, my PR team called to say they had forty-four interviews scheduled.

"Wow!" I said. "When?"

"Tomorrow," was the answer. I was to have forty-four interviews over the course of six hours as part of a satellite media tour.

That day sucked. But I was committed, and I knew why I was there.

At the end of the day, I asked my PR team how much money had been raised.

They told me. Even though they sounded happy, I felt disheartened. It was hard not to take it personally. There was so much work to do, so many kids to support. Even with Jared's contribution, how were we going to get to $200 million?

My PR team tried to assure me it was not my sole responsibility, but that wasn't how it felt to me. I'm goal oriented by nature, and this was a big goal. No matter what happened, I was going to try my absolute hardest to ensure that we not only got there but surpassed our goal.

Our next training exercise was one I had been dreading: water survival training. We needed to train for the unlikely situation where, in an emergency, our capsule would land at an unsupported site somewhere in the world, and we would have to get out of the capsule and into a life raft.

I worried I would fall or get hurt during the egress exercises—but it wasn't because I feared the injury itself at this point. I just knew that if I got hurt, I wouldn't be able to go to space. We were less than two months out, so close but still so far.

It was a hot, late-July day in Cape Canaveral when we got into a training capsule wearing wet suits and motorcycle helmets, since we couldn't practice in the salt water in our space suits. Professional divers were stationed around us, watching out for dangerous Florida marine life.

Our instructions were to jump from the side hatch of the capsule into the hollow bay at the base of the vessel. From that chamber, we would jump into the water to swim to the life raft. I had trouble reaching the step down from the door because of my short stature, so Chris helped to lower me until my foot felt the step.

Each task was increasingly difficult.

Since practice was in the port's calm waters, for the final exercise someone got on top of the capsule and rocked it back and forth, simulating strong ocean waves. We jumped out of the capsule and entered the life raft. We all survived the simulation and were rewarded with Popsicles, which melted all over us while we rested on the recovery ship.

Back at SpaceX, our next training exercise was going to be intense: a thirty-hour simulation. I was slightly dreading this one too, even with my new seat installed.

In space, our capsule would suddenly become a lot more spacious, because microgravity would allow us to float up to the ceiling. But hanging out on the ceiling wasn't possible on Earth, which meant thirty hours straight in a tight capsule with three other people.

From the moment it began, we were busy simulating launch and life in space: We packed and stowed cargo, had our meals, performed research, slept, and did media calls.

Of course, we took the simulations seriously, since we were being thrown problems that, in space, could be life-or-death scenarios. We had to work through a multitude of situations throughout the simulated mission, including that of a sick crew member, where, as the medical officer, I got to pretend-administer a shot.

We also had fun with it.

"Alien sighting," Jared would report to mission control.

"Our poor trainers," I would say, shaking my head and laughing.

Our main crew trainers could hear our internal conversations throughout the thirty-hour simulation. I mentioned several times during conversations with my crew that I would just *love* an iced latte when we finally got out of there.

We were heading off to fighter jet training after that long simulation, so our trainers had uploaded *Top Gun* to our training tablets. The night we slept in the simulator, all four of us pressed play at the same time and we watched the movie together. It was my first time watching it. I finally understood all the *Top Gun* references the pilots on our team had been making.

The movie hit a major plot twist and I sat up to share my shock with my crew, but they were all fast asleep.

My crew and I had so much fun together during that thirty-hour simulation. I laughed so hard my abs were sore.

"We're so funny in space," I said. "I can't wait for the real thing."

We were having so much fun and didn't want it to end. It made me even more excited to spend three days in orbit with these folks. The simulations together also built a great deal of confidence in our crew members, mission control, and our own skills and readiness for space. We all were challenged in several different ways, but we worked well as a team and at the end of the day made it home safely.

Afterward, our trainer noted that instead of saying things like "Five more hours till we're out of here," they noticed us saying, "Aw, only five hours left," which we all thought was a good sign.

The moment we got out of the simulation, someone handed me an iced latte, which I downed during our debriefing. There were definitely upsides to having our training crew listening in.

We flew to Montana that evening for fighter jet training. We spent all of the next day in the hangar, where Jared had a fleet of fighter jets lined up. As an added bonus, our family members were invited to join us. The crew wasn't scheduled to fly until the evening, and in the meantime, he and some of the other pilots on the extended Inspiration4 team took our families for a ride.

Hayden said it was one of the best experiences of his life.

Before the crew took our flights, Jared gave us a briefing, reminding us to take this seriously. For our family, it was all in good fun. For us, this was training. We were going to hit high G-forces and it was going to be uncomfortable. We had to learn to ride it out.

"When we're launching," Jared told us, "we can't just pause it; we have to endure it."

I had my own prebrief with Kidd, who was going to be my pilot. He was a former fighter pilot and Thunderbird, and I knew I was in good hands.

"I want to hit high Gs, I want to do rolls, and I want to take control of the stick."

He asked me how many Gs I wanted to hit, and I said as many as possible. Really, I wanted to hit more than the six and a half that Hayden had hit during his fighter jet flight.

Kidd nodded and said he'd see what he could do.

We took a quick class on ways to eject from the jet in an emergency situation. It did not sound fun. I said a quick prayer that we wouldn't have to eject.

I put my G suit on top of my black flight suit. The G suit would be connected to the jet. Its purpose was to prevent G-LOC, or G-induced loss of consciousness. When I was experiencing high Gs, the suit would squeeze my legs and abdomen to keep the blood from accumulating in those areas and depriving my brain of blood.

To further prevent G-LOC, we also learned breathing techniques and how to squeeze our lower extremities when we were hitting high Gs. I also donned a helmet, gloves, earplugs, and a breathing mask.

No lie, I looked badass.

Kidd helped me into the back seat of the jet, which is not easy to climb into when you're five foot two. With ejection still on my mind, I asked him if he'd ever had to eject before.

"No, don't say that," Kidd replied.

"I'm just asking!"

We lined up on the runway and he asked me if I was ready to go.

"Let's do this," I said.

We took off and the jet glided into the air.

This isn't so bad, I thought.

We met up with the other jets, and then Kidd flew to an area where we had more space to have some fun.

"Ready to pull some Gs, G-Monster?"

"Born ready," I said.

He pulled back the stick and we went straight up, gravity intensifying against us. My G suit wasn't working, so I tightened my legs and abdomen as hard as I could and breathed as I had been instructed.

"Eight Gs," Kidd called out.

YES!

We did barrel rolls so smooth I could hardly tell we were momentarily upside down, until I saw the horizon dance in front and all around me.

Kidd granted my wish and let me take control of the plane, and in my excitement, I yanked the stick. The nose of the plane jerked up in response, and I didn't have control for long.

We met up with the other jets and flew in formation into the sunset. We had been going nonstop for months and I was exhausted. This was a chance to have fun and unwind.

The colors around and below us were stunningly vibrant. We glided in and out of the clouds, and I felt at peace.

From the get-go, I had been very aware of everyone's call signs. Kidd's call sign was Kidd, Leif's was Leif, Jared's was Rook. Call signs are the nicknames that are given to military pilots and astronauts, names that often arise from a joke (usually at the person's expense). Your call sign, once you get one, is with you for life.

There's what you think it means, and then there's often an alternate, less appropriate meaning.

From the beginning, Jared had warned us that we would all be getting call signs later on in training. Now it was time for our naming ceremony.

The crowd at our naming ceremony was made up of the extended Inspiration4 team, our families, and SpaceX employees. Each of us took our turn in the seat at the front of the room. Members of the crowd stood up and shared their ideas, complete with pitches for their chosen names.

I watched and laughed as Chris became Hanks, like Tom Hanks, because Chris did a great job of acting like a sick crew member. Sian became Leo after Leonardo da Vinci, because she's an artist with many talents. When it was my turn, I was highly amused by all of the options that were floating around. The contenders:

Chihuahua, because I'm small but ferocious
Miley, because my go-to karaoke song is "Wrecking Ball" by Miley Cyrus
Comet, as in Halley's comet

At the end, one of the extended Inspiration4 team members, call sign Slick, said I should be named Nova.

"We could call you something easy like Comet," he said, "but your light doesn't flash by every now and again like a comet. It shines even brighter. Like a supernova."

The alternate meaning of Nova, he said, would be No Ordinary Vixen Astronaut.

The crowd voted overwhelmingly for Nova, and it stuck. So now I'm Nova. A new name for a bright and unexpected future. I took the traditional shot of whiskey with my team to celebrate. Then I threw up on the floor.

CHAPTER 16

Launch Week

Quarantining with my family and crew in Florida during launch week reminded me of living in the sorority house and having all of my closest friends within reach at all times. We were bunking at a complex near Kennedy Space Center, just me and the crew and our quarantine friends and family and our SpaceX support team and the documentary crew.

There was one thing holding me back from completely enjoying myself. It was the call with the St. Jude patients that we were supposed to have in space. We were getting close, and nobody was giving me the plan. I was frustrated that I still had no information, and I was worried the St. Jude families wouldn't know about the call because they hadn't been told. All I wanted was reassurance that it was on and details about who and when and what.

I again emailed my contacts at St. Jude, asking for their specific plan of how the call would take place. One of the girls I had been working with throughout the year called me back.

"I'm so sorry," she said, and told me that they couldn't make a live call happen with all the patients.

I felt anger surge through my body. I knew how much it would mean to kids going through cancer treatment to see a survivor in space. That would have meant so much to me while I was fighting.

"I want the name of the person who said no to this call," I said. I'm not a confrontational person by nature, and I don't always speak my mind, but in that moment, it wasn't about me. It was about the kids. "How do they sleep at night, robbing kids with cancer of this experience?"

She was patient and apologetic. "I know how disappointed you must be," she said.

She had no idea.

"Did you all forget I'm risking my life? And this is the *one* thing I asked for." I knew I was being dramatic, but I was caught in the rush. This wasn't just a trip to space for me. It was a mission with a purpose, and that purpose was these kids.

"I'll make some calls," she said. "I'll see what we can do."

To her credit, and with all my gratitude, she did exactly that. A few hours later she called back and said they had a plan to make the call happen. "We're calling this the MacGyver method," she said. "We have a plan to use multiple videoconferencing platforms at once. We think it's going to work."

"Thank you," I said, grateful and gushing. "Thank you so much!"

With that resolved, I could relax and enjoy the pleasures of a week with people I cared about. I was sharing a condo with Sian. My mom and Hayden and Liz had their own condo. I had imagined being so nervous during the days leading up to launch, but it turned out to be a blast.

After a week at home in isolation, cramming for space like I was getting ready for finals, jamming as much extra information into my brain as possible, it was amazing to just be there. Kidd and Leif were there, and so were Anil and Haley. I'd become so close with all of these people, and I loved being able to walk down the hall to see them any time of the day.

I loved getting to know my crew's family and friends on a deeper level. Jared's and Chris's kids were always running around playing and making crafts. The SpaceX family support team provided our meals and we all ate together, then hung out and played card games and drank wine. It felt like college all over again.

September 10, 2021, just days before liftoff and my dad's fourth heavenly birthday. I was looking online for a bananas Foster recipe. I always try to think of ways to honor Dad on these anniversaries, like going to his favorite restaurant or eating his favorite meal or buying an especially beautiful bouquet of flowers, which he loved. My dad didn't cook often, but when he did, he cooked amazingly, especially the steaks he made for us every Sunday. On special occasions he would make bananas Foster, a Louisiana specialty when it comes to dessert.

Dad had taught me his bananas Foster recipe years ago, and now I was kicking myself that I never wrote it down. Though the details were fuzzy, I found a recipe online that was close to Dad's. Since I was quarantining and couldn't go out myself, with the help of the SpaceX employees who had become good friends, ingredients were procured: butter, brown sugar, and cinnamon. We fried bananas, scooped vanilla ice cream into bowls, and covered it all with the warm, delicious, sweet sauce. Then we passed bowls around to our entire quarantined crew.

The group toasted, "To Howard."

"It tastes just like Dad's," said Mom and Hayden, dragging their spoons through their last bites. It was the biggest compliment they could have given me.

I was experiencing what it felt like to heal through time, especially with the right support and having something really big to look forward to.

It was the first time I didn't cry on Dad's birthday.

Two nights before launch, we drove to the ocean. While everyone else was mingling, for a few rare moments I found myself alone. I stared out at the ocean, seeing the reflection of the moon on the water, hearing the roar of the waves. I took a deep breath to smell the salty air and thought about how much I loved being alive. In an interview a few months prior, a reporter had asked me, "Do you even *understand* that space is dangerous?" Yes. I was very aware. It seemed so obvious that there was inherent risk that I found myself continually surprised by how often I was being asked if I understood the danger.

I knew that I was fortunate to have a family who would support my decision, and I knew there was a chance after I said goodbye to my family that I would never see them again. It was a chance I was willing to take because the benefits outweighed the risks. This mission was bigger than me, and the life-changing perspective I would personally gain from seeing the Earth from space would enrich my life in a way that made signing the papers a no-brainer for me.

I trusted that our mission was going to go well, but it wasn't guaranteed. I accepted the risk of death. But oh, did I want to live!

The primary force driving the decision of when to launch was the weather. A 3:00 A.M. launch had been a possibility. Then, a few days beforehand, it was decided that we would aim for 8:00 P.M. on September 15. Personally, I was much more excited about a sunset launch than about launching in the wee hours of the morning.

Once we arrived in space, a little after 8:00 P.M., we would need to be awake for about eight hours to get all our tasks done and settle into our orbital routine. Anil led us in sleep shifting. Each night, we stayed up later and later until we were awake until 5:00 A.M. each night and then slept until around 1:00 P.M. My family tried to stay up with me as late as possible each night but couldn't get anywhere close to dawn.

During the quiet hours of the late nights, my crewmates and I would do last-minute studying, together or by ourselves. One night, we headed to a bay for a midnight kayak trip. Another night, we went on a run around our launchpad. (Well, the boys ran. Sian and I played music and danced as we walked.)

As we neared launch day, one night at 3:00 A.M. we went to visit our spacecraft while it was still in the hangar. The team working on refurbishing our Dragon capsule was still working day and night, finalizing last-minute details before rollout—the dramatic moment when the spacecraft would be transported from the hangar to the launchpad.

At this point the capsule was mated to the Falcon 9 booster and lying sideways. We walked up a ladder to a raised platform where we could closely examine the capsule, trying to stay out of the way of the people working.

"Can I touch it?" I asked.

"Um, okay, you can touch it with one finger," said Maddie, our sweet and funny Dragon integration engineer. So I did. The newly applied white paint felt thick and rough beneath my finger.

The next day, we gathered in a nearby field with our families to watch rollout. The hangar doors opened slowly. We saw our spacecraft emerging. We all screamed. Mom and Liz grabbed their phones to take pictures.

"That's my rocket!" I said, suspended in a state of disbelief, even though I knew it was real. Even though we had had that close-up contact with the spacecraft, there was still a feeling of awe and mystery around it. Seeing it at a distance was almost more impressive.

I looked up at the sky above me, wondering what I would find there.

We got to have another view of the rocket the next day, this time from above because Jared had arranged for a fighter jet flyover. From my seat in the jet, I looked down at the rocket sitting on the launchpad. It was from this spot that I would be lifting off the planet. I was awed and stunned by it all.

L minus three days. Dress rehearsal was a chance for the crew, families, and operators to run through the whole day of launch, exactly as it would be on the day of, and address any concerns if they arose.

We started the day by eating a practice breakfast with our families, as we would on the big day. We walked out and waved to an imaginary crowd. Our families waved us off as we climbed into the Teslas that would transport us to the suit-up room.

I rolled down the window and saw my mom, Hayden, and Liz smiling at me.

"Goodbye," they called out. "Have fun in space!"

They were smiling so big, but I knew this couldn't be easy for them. I felt so guilty for putting them through the stress of watching me launch into space. But I also knew they wouldn't have it any other way.

My vision blurred as my eyes welled with tears. I've never been good at goodbyes, apparently even fake goodbyes, and in three days I was going to tell them goodbye for real. There was a chance that it would be the last time I'd ever see them, these people I loved so much.

When I got back to the complex, I talked to my mom about why I had gotten so upset.

"I feel so guilty for putting y'all through all this stress." I said.

"You do not need to feel guilty," she assured me. "I feel good. I feel everyone praying for me. I feel God's peace that surpasses all understanding."

L minus one day. That evening, we were allowed to bring our families to the launchpad. I hugged my mom as I looked up at the rocket, still struck by how tall it was. I squeezed her tighter. We went up the tower and I showed them the crew access arm—the long hallway I would be walking down the next day to leave the planet. I looked at my phone. It was exactly 8:00 P.M.

"Twenty-four hours, guys. In twenty-four hours, I'm going to be launching into space."

Our last night on Earth.

The Inspiration4 team gathered to cook burgers on the grill outside at our housing complex. I sat next to Leif at one of the picnic tables for a few quiet moments.

A few nights before, I'd had a conversation with him.

"Do you think this is safe?" I had asked him one last time. Leif wasn't just our mission director; he was also an aerospace engineer with a great deal of experience.

"Yes," he said, and went on to explain that everything in life carries risk. You have to decide for yourself if the potential benefits outweigh the risk.

Now it was a clear night and the stars shone above us. I felt excited and at peace about what was to come.

"Look up," he said. "Tomorrow you're going to be looking down."

I knew I would remember this moment forever.

Toward the end of the night, I was able to spend some time with just my family. We played our favorite board game, Sequence, around the table in my mom's apartment. I appreciated that they didn't let me win even though I was leaving the planet the next day.

After a while they couldn't stay awake any longer and hugged me good night. Before I went to my room, my mom sang me a song that she made up, to the same tune as songs she would sing to me as a child.

As I was going to sleep, I thought about the letters.

The day before leaving for launch week in Cape Canaveral, I had been sitting at my kitchen table with a stack of stationery, trying to write letters to my family. It was something we had been advised to do by former astronauts: write letters to our loved ones in case things didn't go well.

I couldn't bring myself to say goodbye just yet, so I began by writing celebratory letters for my family to read after launch.

"Mom, breathe! I made it to orbit!" I started writing to Mom. "I can't wait for our reunion hug. But please don't squeeze me too hard."

Once those letters were written out, ready to be given to SpaceX support so they could pass them along when I was in orbit, I knew it was time to write the goodbye letters. I couldn't put it off any longer. I took a deep breath and told my family how much I loved them. If things went poorly, I knew Hayden would feel especially guilty. This was his field, and he was the one I had called to ask if it was safe to go to space. I told them I was sorry I was gone, that I didn't regret going on the mission, and that they shouldn't have any guilt about letting me go because they couldn't have changed my mind anyway. I asked them to keep sharing memories of me forever and to keep my spirit alive.

Those letters were waiting on my dresser back in Memphis, just in case.

I fell asleep thinking about those letters, hoping my family would never have to read them.

CHAPTER 17

Countdown to Launch

"I'm going to space today," I told my reflection.

I still had to brush my teeth like it was any other day. I still had to get dressed. I started putting on my makeup, completely calm, wondering when I would start feeling nervous. You're supposed to be nervous about going to space.

Haley and Anil came into our condo dancing to Cardi B. I had joked to them weeks before that I wouldn't launch unless they did that, and I had been right to do so, because their performance was very motivating.

Anil took us each into a separate room for medical checkouts. My blood pressure was the highest I'd ever seen it.

"I don't *feel* nervous!" I told Anil.

He assured me high blood pressure before launching off the planet is not unexpected.

Haley braded my hair into two pigtails. I'd had multiple discussions with the space suit engineers months earlier, and we had agreed that this was the best way to manage my long, thick hair in a space suit.

When we met up with our families in the parking garage, mine were wearing their matching powder-blue "Hayley's Ground Krewe" T-shirts. Chris wasn't ready yet, so the rest of us played music loudly and danced in the parking lot. The rest of the day was just like we had practiced during our dress rehearsal.

We sat with our families and ate what they called "breakfast," though it was around 3:00 P.M. My crewmates opted for a light meal; Sian had a smoothie, Jared had a cup of coffee, and Chris had a few bites of food. My plate was stacked high with steak, avocado toast, and a pastry.

"Any last advice?" I asked my rocket scientist brother.

"There are more clouds than you think," he responded.

I gave my mom, Hayden, and Liz each a quick hug goodbye, not wanting to have anything drawn out or get too emotional.

This time, the crowd waiting for us was real, made up of media, SpaceX employees, and our invited launch guests. I spotted my best friends in the crowd, screaming my name in their matching "Hayley's Ground Krewe" T-shirts. My face lit up. I waved and blew them kisses and shouted that I loved them.

We were loaded into the Teslas, and my family came up to the window.

"I'm *so* excited for you, Hayley!" Hayden shouted, visibly elated.

"See you in three days!" Mom said.

There were no tears. We were feeling great.

My crew and I arrived at the suit-up room, where our space suits were waiting. Shit was getting real.

Maria was there, my space suit lying on the table before her. I was so glad to see her. She had fitted me for my space suit nine months earlier, and after several more very enjoyable sessions with her, I had become attached and had requested that she suit me up on launch day.

Maria helped me dive into my space suit and close the zippers. The energy in the suit-up room was palpable; loud music played, and Sian and I moved around the room showing off our signature dance moves.

Before we left, Maria gave me a hug.

"Godspeed," she said.

And with that, we were on our way to the launchpad.

The afternoon was warm and sunny, and the rocket sat tall on the pad. At the very top was our Dragon capsule. As a crew, we stood together at the base of the rocket. Our space suits kept us from looking up with ease, so in order to get the full view, we had to hold on to each other and lean backward together to appreciate the full view.

Spirits were high. We were ready.

At the launchpad, the rocket was sitting next to a large tower with an elevator that carried us up to the deck where the crew access arm was. From there, we would enter our spacecraft. But first, we were able to use the old telephone that sits on the tower. It's the very same phone that has been used for decades by countless astronauts to make their final phone calls to say goodbye to their families before they head into the unknown. The phone has large buttons to accommodate space suit gloves.

My mom answered, putting me on speakerphone with Hayden and Liz.

"I'm calling you from the launchpad!" I told her we were feeling great and excited.

"I'm so excited for you and so proud of you," she said.

"Bye, love y'all!" I said, and let go of whatever was tethering me to Earth. I turned to face the rocket.

I was the first to climb into the capsule, and Anil was there to help strap me in.

"How's your leg feeling?" he asked.

"Good!" I assured him. This was 100 percent true. I wasn't taking any chances this time. The other three crew members climbed inside and got strapped in. We performed communication checks with the ground.

"Nova, how do you hear me?" I heard from mission control in Hawthorne, California.

"Loud and clear."

Maddie served as our Dragon closeout lead. She was the last face we saw on Earth.

"Bye, have fun in space!" she said.

I waved back at her, fanning the air with my large glove, before she climbed out of the capsule and the team closed the hatch.

The interior of Dragon was all clean white, with four seats in the middle. I was in seat number one, at the far left of the capsule. Beneath our seats were three rows of additional cargo and some free space.

Dragon has two windows, slightly larger than airplane windows. There was a door, the side hatch, directly across from the seats. It was from this hatch that we had entered the vehicle and from which we would exit in three or so days (in the Atlantic Ocean).

There was another hatch we were very much looking forward to using on our journey, the forward hatch. Through that hatch was the cupola, a large dome window, the largest window ever flown in space, with room to hang out, the coolest window seat in the galaxy. It was brand-new and had been created for our mission. When they first told us about it, the SpaceX engineers had been just as excited about it as we were. Soon, soon, we would be in there. But not until we had all adjusted to being in space. For now, that door would remain closed.

The walls of the capsule were made up of panels. Behind each panel were cargo slots, storing multiple gray bags belted together that held all of the things we would need: clothing, toiletries, mementos, water bottles, food, medical supplies, research equipment. The plan was to spend three days in space. Since conditions needed to be just right in order for us to reenter Earth's atmosphere, we had enough food and supplies to hang out for five days if we had to. We had our shelf-stable food available, but we also brought a fancy version of an ice chest to space, lined with frozen pouches of coffee that served as the ice packs. Our food was kept in separate metal boxes, each labeled with our seat number, the day, and the meal.

There was a method to all of the organization, one that we had been cautioned multiple times to be mindful of. We would have to be very particular about putting everything back where it belonged, as well as being cautious not to move too much weight around the capsule. We received plenty of lectures about how important proper center of gravity was, and how if it was significantly off, that could have detrimental consequences during reentry.

Chris was in charge of cargo and would be communicating with mission control several times a day to tell them how much we had eaten, how much water we had drunk, and from which compartments we had grabbed the water. As the medical officer, I would lead our daily private medical conferences with our flight surgeon on the ground, as well as address any medical issues that arose in orbit. I'd also be taking the lead with media events, staging the camera and crew, and briefing the crew with our plan for each of our live events in space.

Jared would be monitoring the spacecraft, and he and Sian would be commanding specific procedures. For the most part, our Dragon spacecraft would drive itself, but they had been extensively trained to take control of the vehicle in the event of malfunctions.

T minus 2.5 hours. I had a satchel around my left thigh packed with things I might need, with an iPad on top that I was using to read through procedures one last time. Hidden in the satchel was also a stuffed dog that looked like the golden retriever therapy dogs at St. Jude. It wore a silver helmet and a white space suit that had four silver stars on the chest.

There is a tradition among astronauts that when you reach orbit, you toss out a zero-gravity indicator, usually in the shape of a stuffed animal. As tradition goes, if it floats, you know you're in space. This would be my job when we reached orbit. The zero-gravity indicator is also a surprise, so spectators do not know what it will be until the crew throws it toward the camera in orbit.

Sian played *Star Wars* music from her iPad and we all laughed.

"I spy something white!" I said to my crewmates.

The whole interior of the capsule was white, so they were not interested in playing.

It felt like any other day in our capsule simulator at SpaceX, so I kept reminding myself that this time I was actually on top of a rocket.

I kept waiting for the nerves to hit, but all was calm.

T minus forty minutes. Launch mode. We put our visors down, because from here on out, we could be propelled off the rocket at any minute.

Jared activated the launch escape system, which would remain active until we got to orbit. If the Falcon 9 rocket were to have a problem during propellant loading or launch itself and be at risk of exploding, the Dragon capsule would propel itself away from the rocket through the launch escape system.

T minus five minutes. "Let's fucking do this," I said.

T minus forty-five seconds. "Inspiration4 is go for launch," Jared said to mission control. "Punch it, SpaceX!"

T minus seven seconds. I could hear mission control counting down from ten. We knew there was a two- to three-second delay, so we would launch a few seconds before they hit zero.

Jared could see the real-time count on his screen and was giving us his own countdown.

"Get ready! We're doing this!" he called.

LAUNCH. I felt a large jolt. The increased G-forces came on quickly, but they were relatively easy to handle. From outside I could hear the fastest whooshing sound I'd ever heard.

"We are going FAST!" I yelled. I had a huge smile on my face. This was fun.

About a minute in, the rocket's engines started to throttle down in order to reduce aerodynamic stress; we were entering max-Q, the part of the launch with the maximum aerodynamic pressure on the vehicle. This was the most dangerous part.

I could see Jared's screen from my seat to his left and watched the clock and the G meter. The engines throttled back up to full power.

"We made it through max-Q!" I said. We fist-bumped.

After another minute, the main engines cut off and we were thrown forward in our seats, hanging by our straps.

Main engine cutoff is an expected phase: The nine engines from the first stage of the Falcon 9 rocket shut off so the engine of the second stage can start.

We felt a jolt as the first-stage separated.

The seconds felt long as we continued to hang there. We waited to feel the shift.

The second-stage engine had to ignite, it just *had* to.

The ignition pushed us back into our seats and we cheered and bumped fists. The second stage brought higher-intensity Gs.

"I can't believe we're LAUNCHING right now!" I yelled out.

It was just so smooth. Second stage felt a little bumpier, but just a little. I couldn't believe how smooth it was.

A few more minutes on the second stage, and we were almost there.

I hung in that moment, feeling a sense of safety, feeling Dad there with me.

Before launch, so many people had told me that I'd be closer to my dad than ever when I was in space. That isn't how I viewed it. I didn't need to travel to space to be close to Dad, because he was with me all the time.

He was with me now, and I felt him there, keeping me safe. It was such a profound moment, and I felt it so completely that my eyes filled with tears. The tears shot back like torpedoes into my hair because of the high G-forces.

All of a sudden, there was total silence, perfect stillness, and my straps began to float. I lifted my arms and relaxed them, and they remained in front of me, floating.

We were still restrained in our seats. Another jolt indicated the second stage of the Falcon 9 had been kicked away. Around us, Dragon was coming alive. The fans whirled on and the silence of space was disrupted as our spacecraft began performing the necessary burns to get us to our destined altitude and orbit.

I pulled the golden retriever zero-gravity indicator out of my satchel and let it go. Leashed by a string, it floated in front of me.

You know what they say. When it floats, you know you're in space.

We were in orbit.

CHAPTER 18

Cold Pizza for Dinner (and Breakfast)

Through the window, all I could see was darkness. Then dawn broke through, and pastel blues, purples, and white swirled into my line of sight.

"Is that the Earth?" It wasn't really a question. It was just so unbelievable that I had to verbalize it.

"Yes," said Jared.

"Wow."

But there were things to do aside from feeling awe and wonder. A time line had been uploaded to our iPads prior to launch, and Jared kept a close eye on our progress, calling out regular updates to keep us on schedule. We waited for the cue to unbuckle.

I was the first to release myself from the restraints; I undid my seat belt and allowed myself to rise. My body floated in the weightless environment of space.

This new sensation took some getting used to. At first, my helmet knocked all over the capsule. I went to open a cargo compartment, and as I pushed on it, I was pushed back in return. Newton's third law—for every action there is an equal and opposite reaction—is very pronounced in microgravity. I quickly learned how important it was to anchor myself, usually with my feet, before applying any force.

It was more than an hour before SpaceX gave us the green light to doff, or take off, our space suits. Sian and I went first. (The boys always let us go first.)

We were using the buddy system to help each other out of the suits, which weigh around twenty-five pounds on Earth. I undid the zippers and freed my hands and legs. Then Jared grabbed my helmet and I wiggled myself out from the bottom.

My space suit stayed suspended in the air.

After we doffed our space suits and the long-sleeved shirt and tights we wore underneath, I put on my space clothes. In our meetings with the engineer in charge of our on-orbit clothing, we had each chosen clothes that made us feel comfortable. I picked short-sleeved and long-sleeved T-shirts, as well as leggings and long socks and a sweatshirt. My space gear was sleek, black, and fitted, with our Inspiration4 logo on the sleeve of my top and my name embroidered on the front: Nova Arceneaux.

I also grabbed from my satchel my one-dollar ring I'd bought in Peru all those years ago. I wore it every day on Earth, and three days in space would be no exception.

Two of my crewmates began complaining about nausea, and I sprang into action. I had trained for this situation in particular. Nausea in space is actually pretty common, so much so that it carries a name: space motion sickness. In an attempt to prevent this, we had worn antinausea patches beneath our space suits and had additional meds available in our satchels, but my crewmates needed more.

I was ready for my medical officer's hero moment.

After discussing it with our flight surgeon in mission control, we decided that I would give my crewmates shots of antinausea medication.

"Oooh, I never get to give shots!" I said, very excited, since PAs rarely give intramuscular injections.

"Don't tell me that!" one crew member replied.

"I mean, I perform spinal taps. I've got this."

Drawing up the medication in microgravity was quite difficult, because the medication I needed to draw up was floating around in the vial. I tried techniques I had learned to combat this, such as centrifuging the vial, with little success. Eventually I drew up the medication along with several bubbles.

One of my crewmates expressed concern over the bubbles in the syringe.

"It's going into your muscle; you won't die," I assured them.

I anchored my feet and each crew member anchored themselves. Then I stabbed them in turn and administered the medication. Both were nausea free after the injections.

The mission was saved.

Within hours of our reaching orbit, an alarm went off, one of the big scary alarms, indicating that smoke was detected in the capsule. I rushed over to my seat to grab my bag, which was positioned next to it and contained a breathing mask. I knew this sound well from training. There were three tones of alarm based on the severity of the situation, and this one was the most severe.

Mission control called. According to them, one of the three smoke detectors had gone off, but there was no indication of an actual fire.

We sat in silence. I clutched my breathing mask bag. A fire in orbit would be life-threatening and potentially devastating.

We had spent many hours in training practicing our fire response, which was the most challenging part of training. I thought through the steps now as I sat there. We would each put our breathing mask on and connect it to a place on our seats to receive fresh air. We would then put our fireproof space suits on and receive breathable air from the suits. Then we would have to do an emergency deorbit and come back to Earth, landing at an unsupported landing site. We would potentially sit in the capsule for hours or days waiting for rescue. Or if the capsule was engulfed in flames, we would have to exit the capsule and swim to a life raft. Water survival training would come into play.

Mission control called back confirming that there was no indication of actual fire, but the alarm triggered due to a malfunctioning fan.

I prayed that we wouldn't hear another one of those alarms for the rest of our trip.

We had a few sets of biomedical research to complete while in orbit. On the ground, we'd been trained in the how-to and the why of it all. The main significance of our research would be increased knowledge about how microgravity affects the human body. Astronauts have been participating in medical research for years, but ours was unique and of particular interest to the scientists studying how short-duration space missions would affect regular people.

Astronauts have previously reported a space "fog," or cognitive slowness in space, so we used the iPads to take cognition tests, which were then compared with preflight and postflight tests.

We performed ultrasounds of our large blood vessels to evaluate for fluid shifts in zero gravity. We took ultrasounds of our eyes and bladders, evaluating their size and shape in microgravity. We swabbed several places on our bodies; the swabs would be compared with pre- and postflight samples to evaluate how our microbiome was affected by being in a small capsule with three other people. We did this activity together because it was time-consuming and had several steps. I would call out a body part and we would swab it and store the swab tips in little vials. The most challenging part of this was the trash management. Throughout, this activity created mounds of garbage, and whenever we would open the trash bag to put more in, other bits of trash would try to float out. One annoying part of zero gravity.

We also gave saliva and blood samples. I noticed I was less bleedy in space, likely from the surface tension holding the droplets of blood together on my finger, causing the blood to clot more quickly. With some of the blood we collected, we performed steps to help qualify a diagnostic device that will, hopefully, be used in settings where diagnostic resources are limited. If technology like this is found to be successful, it can be used in more places than just space, like in low-income countries where there are also resource challenges. The blood and saliva samples we collected were later compared with our prespace and postspace samples.

Collecting a saliva sample without gravity was very challenging. When we spit into a tube on Earth, it would fall to the bottom, where it was supposed to go. But in space, the saliva rose away from the tube.

I spat directly into the tube, but the saliva floated out and onto Jared's seat.

"Oops . . . Sorry, Jared!"

We were very forgiving of one another in space.

All of the data collected from our mission was stored in a biobank to be freely shared with aerospace biomedical researchers throughout the world. Nothing like it had existed before, and Inspiration4 was happy to be part of its creation. And through our mission they planned to collect millions of data points in total.

I felt really good about this. As Sian always says, "Solving for space solves for Earth."

Finally, it was dinnertime. The SpaceX medical team had warned us that our appetites would likely decrease in space, since food moves through the gut more slowly without gravity.

I found my "seat 1, day 1 dinner." Inside the box were stacked slices of white pizza with bacon, peppers, and jalapeños. It tasted delicious! I didn't even mind that it was mildly soggy. I stuffed one piece in my mouth, then gave a slice to Chris and asked him to throw it to me so I could try catching it in my mouth.

Epic miss. It sailed right past me.

Jared had a pack of peanut M&M's.

"Hanks! Leo! Nova!" He called us each by name and then lobbed the candies in our direction. All of us missed except for Chris.

Our first throws to one another were hard and fast, since we were used to throwing objects in gravity but it didn't require as much effort in microgravity. When we missed, the candy ended up ping-ponging around the capsule. Newton's third law was again coming into play. The more force we used to throw them, the harder the colorful chocolates bounced off the walls.

It took a moment for us to land on the right technique: the slow and steady throw. Even with the technique mastered, I still wasn't great at catching food. When we missed, we had to track down every piece of candy. If we didn't, it could affect hardware, or even make an unpleasant reappearance during reentry.

Sailing around the capsule chasing candy was so much fun. I *loved* not being bound by gravity. I started doing flips, and I couldn't stop. I curled up in a ball and pushed off, spinning multiple times in a row, until I ran into one of my crewmates or a wall. It was disorienting after spins because I couldn't feel which direction I was facing or where I was in the capsule.

Mission control could see us, and they laughed at me when they called.

I was in heaven. When I had first found out I was going to space, I had so many questions for Hayden. I asked about being upside down in zero gravity, and he cut me off, laughing. "Hayley, there's no upside down in space."

"What? Of course there's an upside down."

I didn't quite understand the concept until I was in space. It's true that in space, being upside down feels exactly the same as being right side up, so any directional statement is arbitrary. Since it all felt the same, I thought, why not be upside down?

I loved every minute of the freeing feeling of floating.

We would all be going to sleep and waking up at the same time, according to the time line. Between the heightened emotions of launch and getting to space, the focus needed for the tasks that we had to complete, and figuring out life in microgravity—including alarm bells and acrobatics—I was so tired.

I brushed my teeth (swallowing the toothpaste, since there was no sink to spit into) and prepared myself for bed. We were sleeping in sleeping bags. Mine was hovering above my seat, and I put my seat belt around it so I wouldn't float off in the night. There was no need for a pillow, since my head wouldn't be touching anything. There are actually zero pressure points.

I was a little concerned about sleeping in space, since I don't always sleep well with people around me. I can never sleep on airplanes. But I was exhausted and I needed to rest. I tried to channel my sleeping habits from home. Before launch I had requested that *Schitt's Creek* be uploaded to the iPad, since I was used to falling asleep to it every night on Earth. I turned on season four, episode one, put in my earpieces, and connected them to my iPad.

I slipped the fabric headband that held my bangs back forward over my eyes. I have slept with an eye mask every night since I was in the hospital at age ten. My grandmother gave me my first eye mask, hoping it would help me shut out the lights of the monitors and the nurses' station.

In the end, I needn't have worried, because I fell asleep within seconds, and I slept more soundly than I usually did on Earth.

I woke up the next morning to find myself levitating, which felt oddly normal. I unbuckled and slipped out of the sleeping bag, pushed off my seat and up into the air, and did a few flips to greet the day. Then I propelled myself over to the coolers and dug out one of the coffee ice packs. That first morning in space, it was still mostly frozen, so I could take only a few sips. The second morning, it was the perfect temperature for a cold brew. That space coffee was just delicious.

I drank from the pouch and looked through the window at the Earth, thinking, *Morning coffee with a view.*

My crew and I had been told we would get really close in space, and it was true. Privacy was somewhat limited. We were forgiving of one another, as we were all rookies in microgravity, which sometimes (often) meant running into one another or accidentally kicking things. But I never got sick of them or felt like I needed a break from them. We all had different roles, and we helped one another and worked as a team. It was so much fun being with them and learning this new environment together.

When the SpaceX psychiatrist interviewed me before we started training, one of the questions she asked was "Do you get claustrophobic?"

I told her all about the time I spent in New Zealand in my twenties living in a camper van with three friends. True, I had never been in a camper van that orbits the Earth at 17,500 miles an hour, but I was confident in my ability to have a great time in tight quarters, and I was right.

Before the flight, we had uploaded a playlist with each of our contributions on it, which meant that we had a great range of genres. One of my selections was "All Star" by Smash Mouth, in homage to my ten-year-old self dancing in the hospital and how far she had come.

We drank water from bottles that had a piece of Velcro attached so we could stick them to other pieces of Velcro around the capsule when we weren't actively drinking from them. I was less thirsty in space and had to actively remind myself to drink water. The bottles were often accidentally kicked, dislodging them from their Velcro attachments. We would float around the capsule tracking them down, quickly becoming familiar with the areas where items were most likely to float off to and get stuck.

We ate all of our meals together, laughing and taking pictures and videos, watching our food float in front of us. We didn't use plates, since the food would have floated right off them. Mealtime could be tricky, what with trying to hold on to the metal container of food (while not letting the other bits of food float out) while also eating, drinking, and holding on to something to keep from floating away.

We shared and traded food like in grade school. Chris offered a Pop-Tart, which I took a few bites of. While I was chewing, he made me laugh, and with my "PAHAHA," the Pop-Tart crumbs shot out of my mouth and formed a cloud that hit Chris and Sian.

"I'm sorry," I said with my mouth full, causing more crumbs to spill out. They laughed as I tried to gather the crumbs floating in the air and put them back in my mouth.

For some reason I was the only person who ended up with bacon, so I shared it with everyone, the only caveat being that they had to catch it in their mouths if they wanted it. We were still perfecting the slow throw, or maybe it was the slow catch that was the issue.

Even though it was fate and a lottery and a social media contest that brought us all together, I couldn't have asked for a better crew to go to space with. It felt like we were always destined to be the crew of Inspiration4.

CHAPTER 19

Around the World in Ninety Minutes

On day two in space, after checking various numbers from the ground, such as the pressure in the cupola, SpaceX gave us a go to open the forward hatch and enter the cupola.

Sian played the theme song of *2001: A Space Odyssey* to add to the drama of the moment. Chris grabbed the handle, turned the lever, and opened the hatch slowly. As he did, I got a glimpse of our bright, colorful Earth.

"WOW!" I said.

"Oh, shit," said one of the guys.

Still, I allowed myself to stare for only a few seconds while the hatch was being opened before shifting into operational mode.

"All right, we've got some work to do," I said, grabbing fabric covers from a cargo bag to cover the lining of the forward hatch.

Then I looked up and the Earth caught my eye again. I could see the entire 360-degree view of the Earth all at once. I could see the whole circle of the globe. I froze. I was suspended, floating, paralyzed by the beauty in front of me.

The Earth is so gigantic, yet so small I could see it all in one view. I finished my task and was the first one to float up into the cupola. Wide-eyed, I couldn't stop staring at the Earth. It was the most beautiful thing I'd ever seen, alive, moving and shifting in front of us.

Around the Earth, we could see the total blackness of space. I was struck by how 3-D the Earth felt. I had seen so many pictures and videos of our planet throughout my life, but actually seeing it from that vantage point, I was able to see the depth of the clouds and the expanse of the curvature. It felt real in a way I had never seen before.

The moon, nearly full during our mission, hung off to the side. Throughout our orbit, the moon danced around, showing up in different places from our perspective.

The bright blue line of atmosphere circled the planet.

"It's so thin," someone said.

Hayden was right. The Earth was all covered in clouds.

Former astronauts had spoken to us about the overview effect, when someone sees the Earth from space, and it changes them and their perspective forever. To be honest, I didn't experience the overview effect on the first day. But when I saw the Earth from the cupola, and could see my entire home planet in front of me, something inside of me changed. I was so shocked by the beauty and the realness of seeing our planet. The Earth was the most beautiful thing I had ever seen, and I felt a strong need to protect her.

How am I getting to experience this? I thought to myself. How was I so fortunate? When I went to space, fewer than seventy women had been to space, fewer than six hundred people total . . . in all of human existence. And I was one of them. The feeling of gratitude was overwhelming. I knew that feeling would stay with me forever.

It was so striking to see landmasses without borders. It felt unifying, but it also made me think of healthcare disparities in a different way. How can someone born on that side of the globe have a completely different prognosis from someone born over here? For example, with childhood cancer, children treated in the United States have an 80 percent chance of survival, but those in low- and middle-income countries have a 20 percent chance of survival. I could see the nations all at once, and it felt more unfair than ever, the ugliness that existed within all of that beauty.

We orbited the Earth about every ninety minutes, so we spent about forty-five minutes in daylight and forty-five minutes in darkness. We moved quickly. Passing the United States took about eight minutes.

Passing over a sunrise or sunset revealed some of the most beautiful scenes. The first sign of sunrise was when the horizon would light up with bright-blue light. The land and sea and clouds below would at first come into focus in pale colors, but the colors would become more brilliant as the light spread overhead.

Sunset was especially beautiful, as the white clouds became mixed with vibrant purples, pinks, and oranges. The clouds would cast long shadows into the darkness that followed. There was a relatively sharp demarcation of darkness, and it was nighttime.

The night passes showed bright city lights. A few times we even saw flashes of bright white lightning as a thunderstorm raged below us. During the night passes we could see stars in the periphery, though I wasn't quite sure if any of them were planets.

Twice I saw a fast-moving and blinking object below. We suspected they were satellites. Sian, the expert geoscientist, and I saw a fuzzy, pale-green light just above the surface of the Earth. She explained it was an aurora.

Based on our schedule, we were often awake when it was daylight over the Pacific region and nighttime over the United States. The oceans were vast, and although we spent the majority of our time over water, we joked that we were always over Australia. It felt like every time we looked out the window and I called out, "Where are we now?" a crew member would float over to our map and laugh.

"Australia again."

Australia was beautiful, a mix of deep reds and neutrals, and from our position we could see the entire continent at once. We even saw Australian wildfires from above, with smoke billowing toward us. I felt so intrigued as I studied the landmass. I have not yet been to Australia, and it made me want to go even more. Near Australia we could see New Zealand, and it brought back wonderful memories of trekking through the North and South Islands with my girls, feeling astounded at some of the most beautiful mountain views and fjords I'd ever seen.

I loved seeing the scattered, small South Pacific islands surrounded by bright turquoise waters.

"Oh, I'm moving these up on my bucket list," I said.

We passed over South America, over Chile and Argentina. The mountain ranges cast shadows on the land below, adding to the depth. I craned my neck, looking for the rainforest farther north. More places I wanted to explore.

And as we passed over North Africa, I saw the brilliant red Sahara desert, with its squiggly lines from windblown sand. I thought of camping in the Sahara right before the pandemic, sitting around a campfire with our Moroccan tour guides as they sang songs in tongues I didn't understand. There's so much to see and explore on our planet. I wanted to book a trip as soon as I got back from space.

Being in space changes you in physical ways as well as emotional ones. The SpaceX medical team had taught us extensively about the physiological changes that occur in microgravity and how our bodies would likely feel when we got to space. Each of us had slightly different symptoms while adjusting to microgravity. These symptoms carried another diagnosis that isn't routinely taught in PA school: space adaptation syndrome.

A few hours after I got to space, I had a moderate headache from fluid that normally is in my lower body shifting to my head in zero gravity. Jared described it as hanging upside down off your bed for hours. I had nasal congestion from the fluid shift, but I could still breathe through my nose, so I didn't take any medications for it.

We had been warned we would likely have back pain, because our spines had been exposed to gravity our entire lives, and suddenly being without gravity caused them to stretch. With the spinal elongation, I likely grew in space, but I didn't have a way to measure my height.

The back pain hit me the second day, and despite the warnings, I was surprised by its intensity. I curled up in a ball, tried to do different maneuvers to help the pain, and took ibuprofen around the clock. The back pain lasted for the duration of the mission, but it did start improving as time went on.

We all got a chance to connect with our families while we were up there.

They could see us on the live internal video stream from SpaceX, and we could hear them over the speakers in our cabin.

When it was my turn, my family shouted questions at me.

"How are you? How's space?" asked Hayden and Liz.

"How's your leg?" asked my mom, always concerned.

"I'm having the time of my life," I told them, talking into the cabin microphone.

I had been so happy to find that my leg didn't hurt at all. On Earth, after a long day, my leg was tired and sore. If I was in one position for too long—like that original space seat—or certain movements stressed the joint, my leg would hurt. In space, I had no pain—without gravity, there was no pressure or weight on my leg, even with all my spins and twirls. It was incredible.

"What have you been eating?" Hayden asked, which was an inside joke because every time we talked to my dad on the phone, he would ask what we were eating.

Then "SPIN! SPIN! SPIN!" they chanted, and I left the microphone suspended in the air as I showed off my zero-gravity spins.

I had a special moment with my dad too. I took his tie out and held it out in front of me. *Here you go, Dad,* I thought. Then I let it go. The colors swirled and whirled as the silky material unfolded itself and danced slowly in the air.

Zero gravity was a source of so much magic for me. When I first got to orbit, the top of my hair was still in braids from being in the space suit. I released the braids, then took off my cloth headband and released my bangs. My long, wavy hair floated around me in all directions. I whipped my head around and the hair followed and bounced. It made me feel like Medusa, but instead of snakes there were long, dirty-blond locks. I was pretty obsessed with my zero-gravity hair.

My crewmates were very patient with my hair, even when it would hit them as I floated around. I didn't realize how much my hair shed until I was in zero gravity and every strand that I lost would wind up floating in the air. I noticed strands sticking to my crewmates' black clothing, and would try to pick them off without anyone noticing.

I knew the real reason I loved my wild hair so much. Losing all of my hair in cancer treatment as a kid was such a horrible experience for me. It was my first realization that I was truly sick. Ever since getting out of treatment, I've kept my hair very long, and I have been so proud of it. It's my symbol of health, of how far I've come.

I took a picture in the cupola with the Earth behind me while holding a picture from when I was in treatment. In that photo, I'm bald, my blue eyes are shining, and I'm wearing a wide smile. I love that picture because even though I was in the height of my treatment, the height of being sick, I was still smiling. I think it's indicative of my attitude throughout treatment. And now here I was, in space with the Earth as my background, holding that photo and showing the world how far I'd come. Smiling as big as ever.

I had packed very carefully to come to space, because I wanted this experience to be as meaningful as possible. I had thought about how to honor the friends I have lost to cancer. I've lost many friends whom I met at St. Jude to cancer through the years, and I've never forgotten them and how much they mean to me. I've kept in touch with their families over the years, and I reached out to say that I wanted to bring their child's picture to space.

When we were in orbit, I took the photos out of the envelope in my bag. One by one, I went through them, seeing my friends' smiling faces. I spent a moment thinking about Luis, who had inspired me to study Spanish and follow my dreams. I felt pangs from missing each friend, but I also smiled, thinking about how good our time together was. Before I wrapped them carefully and put them away, I took pictures of their photos with the Earth in the background, planning to send the pictures to their families when I got back to Earth.

I wanted them to know their children would never be forgotten.

The moment I had been waiting for ever since Jared first told me about our mission was finally here: our space-to-Earth call with St. Jude.

"This is the most important thing we are doing in space," I told my crew. "On this call will be patients in active treatment, survivors, and bereaved families who lost their loved one to cancer."

My crewmates nodded, knowing how much this meant to me.

A few of the patients I was closest to would be asking us questions on that call, I had told them multiple times. We all had our own reasons for being here. This was mine.

Seeing how enthusiastic my crew was meant so much to me.

We positioned ourselves in front of the camera. I was upside down, because I thought it would be fun for the kids watching, and I wanted them to believe I was really in space.

The first thing I heard on the call was my friend Joel's voice introducing me. Joel is a dear friend and fellow osteosarcoma survivor. We met way back when I was a kid in treatment and he was a new St. Jude employee. We had both come so far, and I was so happy to hear his voice.

"Hello, everyone, and welcome to our Dragon capsule here in space!" I said into the cabin microphone. "My name is Hayley Arceneaux, and some of y'all on the call may know me as your PA. I work at St. Jude, I have the best job ever, and like y'all, I was a St. Jude patient as a kid."

The questions were amazing.

"Are there cows on the moooon?" one kid asked.

"What was your training like?"

"What is life like without gravity?"

"Qué hacen para divertirse en el espacio?"

"Are there aliens in space?"

We each took turns answering their questions, and I used my Spanish to answer the question about what we do for fun in space.

Jared helped me with a zero-gravity demonstration by throwing a peanut M&M at me, and I actually caught it in my mouth. *Boom!* On camera. Documented forever.

I floated to the top of the capsule, showing off our view of Earth from the cupola.

And I told them all how I truly felt. I told them how honored we were to have them in our spacecraft with us, and that we were doing it all for them. "I wanted to tell you that I was a little girl going through cancer treatment just like a lot of you," I said. "And if I can do this, you can do this. I'm so proud of each and every one of you. I can't wait to tell you the stories when we get back to Earth."

I wanted our mission to show these kids, and everyone following along, that they could dream big dreams. I wanted to show them that what may seem impossible is possible.

One thing I'm pretty sure is that St. Jude is now full of future astronauts, because they all want to see for themselves if there are actually aliens out there.

CHAPTER 20

Splashdown

Sian and I sat in the cupola, staring at our home planet. We were just hours from heading back to Earth.

"It was a real pleasure to go to space with you," Sian said.

We looked each other in the eye. "You too," I said. She was my sister for life.

I knew that for the whole rest of my life I would miss this feeling of looking at our globe hanging in the black, with the moon scampering alongside. I focused on committing the details to memory, reminding myself to always remember how this view made me feel.

Still, I felt at peace. I wasn't sure if I was ready to go home, but I was looking forward to fresh food. This last day in space I hadn't even eaten; the physiological effects of zero gravity had caught up with me and I wasn't feeling hungry. Plus, I was getting tired of the pizza. For one of the research endeavors, I had worn a continuous glucose monitor to evaluate how glucose fluctuated in microgravity, and I kept an eye on it to make sure my glucose didn't drop too low.

I was so ready for the fried chicken I knew was waiting for me on Earth. I took a few last pictures and last glances, then floated down out of the cupola.

I didn't look back, because it was too hard. I hate goodbyes.

We changed into the clothes we would wear under our space suits and started fluid loading.

To prevent feeling light-headed or fainting when we got back to Earth because of the microgravity fluid shift, we chugged water and took salt tablets to increase the volume in our blood vessels and raise our blood pressure. I spun the water bottle in front of me, watching the bubbles shift around as it spun, then took a gulp and swallowed a salt tablet as I took a few last spins in space. It's not advised to swallow pills upside down on Earth, but since technically there's no upside down in space . . .

We put on our space suits and strapped into our seats. I already missed my favorite sensation in the world, floating in zero gravity. At this point my abs were sore from how many spins I'd done in those three days, but I didn't care. It was worth every sore moment.

Dragon performed a few burns to lower our altitude and, finally, to bring us home. Similar to launch, we strapped in two hours prior to splashdown because of this.

Jared and Chris started watching *Spaceballs*. I watched on Jared's screen for a few minutes, but I was distracted.

"How about we list all the ways we could die during reentry," I said.

My crewmates looked at me.

In retrospect, it was a very morbid thing to bring up, but at the time I thought it would help to mentally prepare myself.

I started off by listing a few potentially catastrophic failures: failure of the thermal protection system, failure of the parachutes to deploy, center of gravity being off and sending Dragon into a tailspin.

Jared added a few more.

I became quiet again. It felt like a lot of things could go wrong. During launch we had the safety net of the launch escape system, but during reentry we had no backup plan.

I watched the clock counting down to splashdown time, the number getting smaller and smaller. Then it was time.

We put the visors of our helmets down. Dragon's burns were rhythmic in my ears. Then I heard a whooshing sound. For the first time in three days, I started feeling weight, and it felt ridiculously heavy.

"Zero point three Gs," Jared called out.

What? This is only one-third of Earth gravity? It felt like I had people lying on top of me.

"And we have to make it to four point five Gs?" I asked.

After three days of no gravity, just that small amount of gravity felt intense. G-forces continued building. I breathed in the way we had been trained and focused on keeping my lungs inflated. I reminded myself it would be over in less than ten minutes.

As we started entering the atmosphere, from the window I could see bright flashes of light that looked like lightning. Then flashes of red fire. The windows fogged from the plasma on the vehicle, and we couldn't see anything else. Our capsule, and all of us inside, shook slightly as we made our way through the atmosphere.

I'm not going to lie; in that moment I felt fear. When I was eleven, my mom woke me up one morning to tell me the space shuttle *Columbia* had broken apart during reentry and all of the astronauts on board had perished. While we were going through reentry, I thought back to that morning.

We were falling, plummeting back to Earth. Our capsule was on fire at about 3,500 degrees Fahrenheit. We were in a communications blackout, unable to talk to mission control. I thought if I died, I might see my dad's face at any moment.

There was nothing I could do about it. I just had to trust and pray.

Communication was restored with SpaceX and we were told to brace for the first set of parachutes.

BOOM! BOOM! BOOM!

We heard what sounded like explosions. And as the parachutes deployed, our capsule was jolted upward. It started shaking and swaying harder than ever.

Jared told us not to celebrate the first set of parachutes. We needed a healthy second set of parachutes to live.

We were told to brace again, then *BOOM!*

The force from the second set of parachutes opening was slightly less dramatic, but the whooshing was louder than ever.

"Four healthy mains," we heard from mission control.

"THANK GOD!" I shouted. I knew at this point we were going to be just fine.

We were in normal Earth gravity as the parachutes glided our capsule down, slowing its velocity. It felt peaceful, and we heard what sounded like a summer breeze outside.

Jared called out how many meters we had until splashdown. We put our arms over our chests in brace position at two hundred meters. He listed off the numbers: "One hundred. Fifty. Twenty-five. Twenty. Fifteen. Ten. Seven. Three. Zero. Umm, negative three. Negative five."

There was no doubting when we hit the water. *BOOM!* It was a more intense impact than we had expected, but it didn't hurt. A wave hit our capsule. We both heard it and felt it jostle us, but we couldn't see through the windows. Our capsule rocked in the water.

"On behalf of SpaceX, welcome back to planet Earth," said mission control, as we cheered and fist-bumped.

"Inspiration4 is mission complete," Jared called back.

We had done it. We were home.

Our capsule was lifted onto the recovery ship, and as the side hatch was opened, Anil entered the capsule.

"Welcome home, earthlings!"

He evaluated each of us medically, assessing the strength of our legs.

"My legs feel really weak. Is this normal?" I asked.

He assured me it was.

I was the first to exit the capsule. Before I did, I stood, and a medical team stood around me. They asked if I felt light-headed. Cautiously I said no.

As I stood there, Kidd came into sight.

"Zero-G-Monster!" he shouted.

"Zero-G-Monster? YES!" I called back. I loved my new nickname.

The medical team assisted me up, and I climbed out the side hatch. I smiled and waved to the camera crew that was waiting and gave them a thumbs-up. I then walked into the medical bay, where I had a nurse waiting for me. I lay on the stretcher.

"I feel drunk," I said. After the short walk to the medical bay, I felt dizzy and weak. She helped me out of the space suit.

"Where's the fried chicken?"

It was waiting for us on the recovery boat, and it was cold, but it was more delicious than I could have imagined. I ate lying down, crumbs covering my shirt while the nurse took my vitals. Above us, the moon still danced around the Earth, and I could see the stars, and I still didn't know which of them were planets.

After we were all evaluated, we were loaded into a helicopter. Our families were waiting for us at Kennedy Space Center.

"OH MY GOODNESS!" my mom screamed when she saw me, squeezing me tight. I squeezed back even tighter.

I was home.

CONCLUSION

I'm Still Me

JANUARY 10, 2022

The summer before I was diagnosed with cancer, twenty years before I was invited to join Inspiration4, when my family and I took a tour of NASA because my little brother was obsessed with space, Mom and Dad and Hayden and I had our picture taken in front of a green screen. It's still somewhere in my mom's house, a souvenir photo in which we are floating together as a family in zero gravity. Of course, we weren't; one of the employees edited it to make it look that way.

It amazes me to think that I got to spend time in a place where I was actually floating, no Photoshop necessary. That experience of weightlessness will be with me forever. That experience of seeing our planet with its constantly shifting cloudscapes, the way the moon moved around her in the blackness, will be with me forever.

So will the feeling I had the evening we landed and found out that we had not only met but surpassed our $200 million fundraising goal for St. Jude. I was overjoyed. I had been able to channel the anger I feel toward cancer—for taking my father and so many of my friends—and do something about it. All that hard work, all those interviews, it was all worth it. We did it. It truly felt like not only mission complete but mission accomplished.

The money we raised will, hopefully, go a long way to helping eradicate cancer. And I am more committed than ever to being a part of this fight until cancer is absolutely a thing of the past.

It was also incredible to learn that despite my fears and worries about the call with St. Jude in orbit, in the end we spoke live with 1,500 families on that call. When I found out that I had been directly communicating with so many kids and families, I couldn't stop smiling. Two decades after we first walked into St. Jude and Ms. Penny said that we were part of the family, it is still true.

I saw Ms. Penny a few days after we returned to Earth at our welcome-home party, and she had the same kind eyes and sweet voice. She told me how proud she was of me, and I thought back to how much she had meant to a scared ten-year-old and her mom all those years ago, and how grateful I am that I can now offer comfort like that to kids and their parents.

My cancer besties Hannah and Katie were there, still healthy and cancer free, living their best lives. Hannah came with her fiancé and Katie was with her husband and two kids. Dr. Doom wasn't at my return party, but he was at launch, and I found out later that he brought the model of my prosthesis to show Anil, he was so proud. (I have a photo of the two of them at a bar, holding the prosthesis up in the air.)

So many people I've known and loved through the years, who have been with me on the toughest days, showed up to celebrate with me on this most joyous day. All my best friends and families were there, including Gabrielle Corsentino and the rest of the Corsentino crew, and some of my favorite co-workers from the emergency department and St. Jude. Lizzie was there too, watching alongside my family. And a person who was recently added to the mix, Mom's new boyfriend, was by her side.

In moments like this, life feels beautiful and full circle.

If I'm being honest with you, and I am, I have to tell you that going to space was never really my dream. It's not something I thought much about. Instead, my whole life has been my dream—just living was my dream. Getting to graduate from high school, to study medicine, to work at St. Jude. Getting to travel, to have adventures, to say yes to things, to keep looking forward. Those were my dreams.

Now my dreams are expanding. In October, a few weeks after our successful mission, we went back to SpaceX and were presented with astronaut wings, a beautiful symbol of an incredible journey. I picked them up for a closer inspection, and the silver Dragon wings glistened in the light.

While I was at SpaceX, Anil invited me to take a walk with him and discuss opportunities to join the SpaceX medical team. He knew I felt passionate about sharing my experience and what I've learned with future fliers. Two months after our mission, I became part of the team helping medically train commercial astronauts. It's not a big commitment of time, so I can continue doing what I have always wanted to do—work with the kids of St. Jude. I feel like the two jobs actually work well together. Both at St. Jude and at SpaceX I went through quite an experience, with treatment and with training. I was supported by both organizations in such a special way when I needed it most, and now I can repay the favor. Now that I'm on the other side, I can give back, share what I learned along the way, and support the current patients and astronauts.

Also, it turns out I love space medicine. Who would have thought?

I walked back into my job at St. Jude on January 7, 2022, just about one year after I first got the call to go to space. I was a little bit nervous and a whole lot excited to be getting back to my dream job. My very first patient of the day, a little boy with big, curious eyes, told me he wanted to be an astronaut, and as it happened, I had some videos on my phone I was happy to share.

My space suit has been donated to St. Jude, and if he wants to visit it, he can. So can all the children and their parents, some on their very first day at the hospital, wrestling with the fear of the unknown. I hope it serves as a symbol of what they can do too, how wonderful life after cancer can be. I'm the first pediatric cancer survivor to go to space, but as I tell my patients, I'm certainly not the last.

Every time I walk onto the St. Jude campus, I see the newest building, which is now named the Inspiration4 Advanced Research Center. I can't believe how proud I am of a building, but really, I'm so proud of what it represents. When our crew was at the building dedication, Jared said it was to be dedicated to all of the children who went to the stars before their time. My hope is that with the research that Inspiration4 funds, more and more children will have the chance to grow up and discover their dreams right here on Earth. Or in space.

There was a time in my life when I didn't think I'd get to turn twenty-one. Incredibly, I not only got to turn thirty, but by the time I did, I had spent 10,947 days on Earth and three days in space. What a wild ride.

I've been asked several times whether, if I had the opportunity, I would tell my younger self going through cancer treatment that she would one day become an astronaut. The truth is, I would not tell ten-year-old me that she would one day go to space. The beauty of life is that you don't know what's going to happen. That's why, even on your most difficult days, you have to hold on to hope that there will be better ones, so great that you can't even imagine.

I still have moments, very frequently, when I think to myself, *I can't* believe *I went to space.* I recognize how fortunate I am to have experienced something so few have been able to see and feel. There are times when I hope I'm not going to just wake up and realize it was all a dream. Becoming an astronaut is not something

I ever imagined could even be possible. It came in like a wrecking ball.

It wasn't an easy year, and it wasn't always easy to stay vulnerable and honest with cameras trained on us at every moment. The result, though, was a documentary that lets the world see what we got to see. The occasional annoyance I felt at constantly being filmed evaporated when I saw the result. When *Countdown: Inspiration4 Mission to Space* was released on Netflix, I was able to relive the best year of my life. Watching it allows me to feel proud of how hard I worked, how brave I was, and how much I learned about myself. So did the experience of writing this book.

One of my first questions when I found out I was going to space was "Are we going to the moon?" I knew basically nothing about space then, and I am amazed how much I learned over the months that followed, all the way to becoming an astronaut. A few months after our mission, I was writing an email. I casually typed the phrase "the time I went to space," then laughed out loud.

Going forward, I plan to continue doing what I love the most: traveling the world. Space was an unbelievable experience, and it made me appreciate Earth in a new way, so that I wanted to get back down to continue exploring on the ground. There's so much to explore on our planet and beauty to see and people to meet and things to learn. I've got a long bucket list, but I think that's how it should be. I'll never stop daydreaming about my next adventure and having something on the books to look forward to. Plus, I owe it to Australia to check it out.

People keep asking if I've changed. Absolutely, yes, I have. I haven't transformed into a different person, though. I've just become more me. This year I've seen my confidence and strength grow. I have pushed past the limits I thought I had and learned more about myself in the process. Going forward, I'll know that the experiences I fear the most can be the most rewarding, and that true toughness means taking uncertain steps exactly as you are.

I believe in saying yes to opportunities that can change your life, even opportunities that scare you. Take the chance and you will feel, and learn, and grow, and become even more you. Following your dreams can take you to dreams you didn't know you had.

Not everyone has had childhood cancer to overcome, but everyone has had something. Something that hurts to the core, makes you doubt what the future will hold, and even makes you lose hope. What I've learned in my life is to not give in to that despair, not to lose hope, no matter what. Happiness and hope are, at the end of the day, choices. And they are worth choosing.

The day after I got back to Memphis, St. Jude threw a parade in my honor. I was loaded onto the back of Dr. Doom's convertible, with his wife at the wheel, wearing her "Hayley's Ground Krewe" T-shirt. The first stop on the parade was in front of the main hospital. Patients and their families flocked the sidewalks, holding pom-poms and homemade posters. The St. Jude CEO had the microphone. He asked if I could have imagined when I walked through those front doors nearly twenty years ago that I would one day be welcomed back to St. Jude as an astronaut with a parade.

The truth is, the dreams I had don't even compare to the glorious, beautiful life I have been given. I don't know what's going to happen, and I can't imagine it, not even a little bit. I'm just so excited to see what comes next.

Acknowledgments

I hope this book shows that I am who I am because of the people who have supported me, lifted me up, and made me laugh through it all. I am beyond grateful to have so much love in my life.

Some particular shout-outs as follows:

Thank you.

To Sandra Bark, for helping make my dream of this book a reality, for understanding who I am and helping me share it with the world. You are the type B to my type A, and early mornings to my late nights, and together we made the perfect team.

To Eliza, Cait, Keren, and the entire team at Convergent, for believing in my story, for your wisdom, and for guiding me through this process.

To Ashley Longshore, for the gift of this cover that I love so much and that embodies how I see myself, a badass and feminine astronaut.

To Mom, for cleaning more vomit than anyone ever should, for being our family's rock, and for being the best friend, travel buddy, and role model. Thank you for always supporting my adventures, even when they make you scared. Because of you, I am me.

To Dad, for teaching me the importance of exploring the world and forming relationships with people along the way. I will always try to live my life in a way that honors your legacy and the deep love you showed our family.

To Hayden and Liz, for answering my innumerable rocket science questions and for always showing up. Hayden, you've been there through it all, and you are my go-to now and forever. Liz, I am so grateful for your loving support through Dad's sickness and launch week and everything in between.

To Aunt Kerryn and Lauren, my extended immediate family. I can't imagine going through life without the two of you.

To my girls, you know who you are. I became who I am with you all by my side. Thanks for being the best, most ride-or-die, supportive Krewe a girl could have.

To my care team at St. Jude, especially Lizzie, Dr. Neel (Doom), and Dr. Jane. So many people worked together to help me not only survive but have an unbelievably positive experience with cancer that made me want to grow up and do what you do. Thank you for carrying me through the hardest days and helping maintain my spirit.

To Jared Isaacman, I'll never be able to express my gratitude for your trust in me, for bringing me to the stars, and for your friendship—as well as for your passion and commitment to ending childhood cancer. You're making the world a better place.

To the rest of my crew, Sian and Chris. The four of us will share such a special, unique bond forever. I couldn't have asked for a better crew. Also . . . thank you for your patience with my zero-gravity hair.

To the entire Inspiration4 team, who became family. We wouldn't have gotten to where we were without your tireless efforts, which were both seen and deeply appreciated. I love you all.

To the SpaceX team, who instilled confidence, kept us safe, and helped give us the most incredible experience.

To the ALSAC/St. Jude team, who believed in me and helped give me the opportunity of a lifetime.

And lastly, to my patients and their families, for being my daily inspiration. Working with you all is the greatest honor of my life.

A Note About the Cover Art

The astronaut with high heels and a disco ball was painted by a Louisiana-based artist named Ashley Longshore. I've been a fan for years and reached out to her when I saw this painting to tell her how much I loved it. The image of the astronaut is exactly how I see myself—badass and strong and feminine. Although the space suit in the painting doesn't look exactly like my SpaceX space suit, it's all about the vibes.

About the Author

Hayley Arceneaux is a physician assistant at St. Jude Children's Research Hospital, a career she committed to at age ten, after surviving pediatric bone cancer at St. Jude. She served as an ambassador for the hospital when she joined the first all-civilian orbital space mission, Inspiration4, in September 2021, and spent three days in orbit. At age twenty-nine, she became the youngest American in space, the first pediatric cancer survivor in space, and the first astronaut with a prosthetic body part. She hails from Louisiana but is now living in Memphis, Tennessee, with her Aussiedoodle, Scarlett. She continues traveling and exploring the beautiful planet she gained a unique perspective on from space. Connect with her on Instagram at @hayleyarc and on Twitter at @ArceneauxHayley.

About the Type

This book was set in Garamond, a typeface originally designed by the Parisian type cutter Claude Garamond (c. 1500–61). This version of Garamond was modeled on a 1592 specimen sheet from the Egenolff-Berner foundry, which was produced from types assumed to have been brought to Frankfurt by the punch cutter Jacques Sabon (c. 1535–80).

Claude Garamond's distinguished romans and italics first appeared in *Opera Ciceronis* in 1543–44. The Garamond types are clear, open, and elegant.